Complement and Infectious Diseases

Author

Douglas P. Fine, M.D.
Associate Professor
Division of Infectious Diseases
Department of Internal Medicine
The University of Texas Medical Branch
Galveston, Texas

CRC Press, Inc.
Boca Raton, Florida

Library of Congress Cataloging in Publication Data

Fine, Douglas P   1943–
  Complement and infectious diseases.

  Bibliography: p.
  Includes index.
  1. Communicable diseases—Immunological aspects.
  2. Complement (Immunology)   I. Title.
  RC112.F494   616.9'0479   80-29396
  ISBN 0-8493-6075-7   AACR1

This book represents information obtained from authentic and highly regarded sources. Reprinted material is quoted with permission, and sources are indicated. A wide variety of references are listed. Every reasonable effort has been made to give reliable data and information, but the author and the publisher cannot assume responsibility for the validity of all materials or for the consequences of their use.

All rights reserved. This book, or any part thereof, may not be reproduced in any form without written consent from the publisher.

Direct all inquiries to CRC Press, 2000 N.W. 24th Street, Boca Raton, Florida, 33431.

© 1981 by CRC Press, Inc.

International Standard Book Number 0-8493-6075-7
Library of Congress Card Number 80-29396
Printed in the United States

FOREWORD

This manuscript originated in my attempts to learn the extensive literature relating to the role of complement in bacterial infections. It seemed at the time that much relevant data was scattered in old, or obscure, or diverse spots; and I suspected that other investigators in this area shared my curiosity as to what exactly has been learned since complement was discovered in the latter part of the 19th century. It soon became clear that studies with microbes other than bacteria were also highly pertinent to any understanding of this area. Thus, my first goal has been to assemble as much of the literature as I could find and to share this organized bibliography. Practicality has limited the scope to English language sources published prior to 1980 and available to me in Galveston. But I have been fortunate to have had access to the Moody Medical Library of The University of Texas Medical Branch; the journal collection is extensive and very few references proved unobtainable.

Secondly, I have tried where possible to synthesize data and interpret what is known without introducing too many of my own biases. I have been especially interested in observations relevant to the role of complement, particularly the alternative pathway, as a basic host defense and the ways virulence may reflect the interactions of microbes with this system. Review of the studies of bacteria and parasites proved especially interesting in this regard.

Finally, this monograph has been written with the notion that it may be read by complementologists relatively untrained in microbiology and infectious diseases and by microbiologists and infectious diseases specialists relatively untrained in complementology. Thus, abbreviations are used sparingly and jargon has hopefully been minimized. Furthermore, I have attempted to offer general background information where it seemed relevant or helpful.

Many people made this book possible. I specifically wish to thank Ms. Donna Johnson, whose typing skills and good humor were indispensable. Ms. Susan Menotti, Bonnie Rhew, and Shirley Saddleback also helped with typing and the multiple other tasks involved. Mr. Mitch Magee prepared the illustrations. Mr. Randy Rumsey kept our lab going and kept people away from my door when I was writing. I am happy to acknowledge my debts to these generous people. I could not have finished this job had my colleagues in the Infectious Diseases Division not encouraged me, put up with my ill humors, and carried part of my clinical and teaching loads. Grateful thanks go to Drs. James A. Reinarz, James C. Guckian, Jon T. Mader, Richard B. Pollard, Michael G. Bullen, and Jo Ellen Schweinle. Dr. Guckian provided helpful reviews of several sections; and I thank Drs. Adam Ewert and Leroy J. Olson for their reviews of the chapter on Parasites; Dr. Reinarz for review of Fungi; and Dr. Pollard for review of Viruses.

Finally, I thank my wife, Gail, and daughters, Laura and Allison, for their loving encouragement and support. Without them, I could not have done this work. They are surely as glad as I to see it finished.

Douglas P. Fine, M.D.

THE AUTHOR

Douglas P. Fine, M.D., is Associate Professor of Internal Medicine at The University of Texas Medical Branch in Galveston.

Dr. Fine received his B.A. from The University of Texas at Austin in 1966 and his M.D. from The University of Texas Medical Branch at Galveston in 1968. After an internship in Internal Medicine at Galveston, he trained as an Assistant Resident in Medicine at Vanderbilt University (1969–1971) and a Fellow in Infectious Diseases at the Nashville Veterans Administration Hospital and Vanderbilt University (1971–1972). Dr. Fine served in the U.S. Army Medical Corps from 1972 to 1975 and was stationed at the U.S. Army Medical Research Institute of Infectious Diseases, Frederick, Maryland. Since 1975, he has been on the faculty at The University of Texas Medical Branch.

Current research interests include opsonization of pneumococci and other bacteria early in infection in the nonimmune host, functional assays for efficacy of pneumococcal vaccines, mechanisms of bacterial opsonization and phagocytosis, and nonopsonic bacterial adherence to human complement receptors. Dr. Fine is the author or coauthor of over 20 published articles and chapters. He is a Fellow of the American College of Physicians and of the Infectious Diseases Society of America; a member of the American Federation for Clinical Research, the American Society for Microbiology, and the American Association of Immunologists; and a Diplomate of the American Board of Internal Medicine, certified in Internal Medicine and Infectious Diseases.

DEDICATION

To my parents, Eldon and Eleanor Fine,
and to my wife, Gail

TABLE OF CONTENTS

Chapter 1
The Complement System 1
- I. The Complement System 1
  - A. Terminology 1
  - B. C3 1
  - C. The Classical Pathway 4
  - D. The Alternative Pathway 4
  - E. Terminal Sequence 6
- II. Biological Roles of C3 Receptors 6
  - A. Opsonization 6
  - B. Role of Complement in Intraphagocytic Killing of Microbes 13
  - C. Other C3 Receptors 14
    1. Erythrocytes 14
    2. B Lymphocytes 14
    3. Renal Glomeruli 15
  - D. Variability Among Complement Receptors 15
- III. Anaphylatoxins, Chemotaxis, and Stimulation of Leukocytes 16
- IV. Complement-Coagulation Interactions 18
- V. Lysis 18
- VI. Complement Methodology 20
  - A. Complement Assays 20
    1. Functional Assays 20
    2. Immunochemical Assays 20
    3. Metabolic Studies 20
  - B. Strategies to Study the Role of Complement in Infectious Diseases 21
  - C. Assessment of Complement Pathways 21
- References 22

Chapter 2
Human Complement Deficiencies 31
- I. Congenital Complement Deficiencies 31
  - A. Classical Pathway Deficiencies 31
  - B. Alternative Pathway Deficiencies 32
  - C. C3 Deficiency 32
  - D. Terminal Component Deficiencies 33
    1. C5 Deficiency 33
    2. C6 Deficiency 34
    3. C7 Deficiency 34
    4. C8 Deficiency 34
    5. Summary 34
  - E. C5 Dysfunction 35
  - F. Treatment 36
- II. Acquired Complement Deficiencies 36
  - A. Sickle Cell Disease 36
  - B. Splenectomy 37
  - C. Protein-Calorie Malnutrition 38
- III. Summary and Comments 38
- References 39

Chapter 3
Bacteria 43
- I. Classification 43
- II. Bacterial Surfaces 43
  - A. Cell Membrane 43
  - B. Cell Wall 43
    - 1. Gram-Positive Bacteria 43
    - 2. Gram-Negative Bacteria 44
  - C. Capsule 45
  - D. Interactions of Host Defenses with Bacterial Surfaces 46
- III. Spirochetes 46
  - A. *Treponema pallidum* 46
- IV. Gram-Negative, Aerobic Rods 47
  - A. *Pseudomonas aeruginosa* 47
  - B. *Francisella tularensis* 48
- V. Gram-Negative, Facultatively Anaerobic Rods 48
  - A. Enterobacteriaceae 48
    - 1. *Escherichia coli* 48
    - 2. Salmonellae 50
    - 3. Shigellae 51
    - 4. *Klebsiella pneumoniae* 51
    - 5. *Serratia marcescens* 52
    - 6. *Proteus mirabilis* 52
    - 7. *Haemophilus influenzae* 52
- VI. Gram-Negative, Anaerobic Bacteria 53
  - A. Bacteroides 53
  - B. Fusobacteria 53
- VII. Gram-Negative Cocci and Coccobacilli (Aerobes) 54
  - A. Neisseria 54
    - 1. *Neisseria gonorrhoeae* 54
    - 2. *Neisseria meningitidis* 55
- VIII. Gram-Positive Cocci 55
  - A. *Staphylococcus epidermidis* 55
  - B. *Staphylococcus aureus* 56
  - C. Streptococci 60
    - 1. Streptococci other than *Streptococcus pneumoniae* 60
    - 2. *Streptococcus pneumoniae* 63
- IX. Gram-Positive, Asporogenous Rod-Shaped Bacteria 66
  - A. *Listeria monocytogenes* 66
  - B. *Erysipelothrix rhusiopathiae* 66
- X. Actinomycetes and Related Organisms 66
  - A. Corynebacteria 66
  - B. Propionibacteria 66
  - C. Mycobacteria 67
    - 1. *Mycobacterium tuberculosis* 67
    - 2. *Mycobacterium leprae* 67
  - D. Micromonosporaceae 67
- XI. The Rickettsias 68
  - A. *Rickettsia rickettsii* 68
- XII. The Mycoplasmas 69
  - A. *Mycoplasma pneumoniae* 69
  - B. *Mycoplasma hominis* 70

	C. Other Mycoplasmas	70
	D. *Ureaplasma urealyticum*	70
	E. *Acholeplasma laidlawii*	70
	F. Comments	71
XIII.	Endotoxin (Bacterial Lipopolysaccharide)	71
References		74

Chapter 4
Fungi 85

I.	Aspergillus	85
II.	Candida	86
III.	*Torulopsis glabrata*	87
IV.	*Cryptococcus neoformans*	87
V.	*Coccidioides immitis*	89
VI.	*Paracoccidioides brasiliensis*	89
VII.	Summary and Comments	89
References		90

Chapter 5
Viruses 93

I.	Poxviruses	97
	A. Variola Virus (Smallpox)	97
	B. Vaccinia Virus	97
II.	Herpes Viruses	98
	A. Herpes Simplex Virus	98
	B. Cytomegalovirus	99
	C. Epstein-Barr Virus	100
	D. Infectious Bovine Rhinotracheitis Virus	101
III.	Papovaviruses	101
IV.	Picornaviruses	101
	A. Enteroviruses	101
	B. Hepatitis A Virus	101
V.	Togaviruses	102
	A. Alphaviruses	102
	B. Flaviviruses	102
	C. Rubella Virus	103
	D. Equine Arteritis Virus	103
VI.	Orthomyxoviruses	103
VII.	Paramyxoviruses	104
	A. Newcastle Disease Virus	104
	B. Measles Virus	104
VIII.	Rhabdoviruses	105
IX.	Retroviruses	105
X.	Arenaviruses	106
XI.	Coronaviruses	107
XII.	Hepatitis B Virus	107
XIII.	Bacteriophages	108
References		109

Chapter 6
Parasites 115

I.	Protozoa	115

	A. Plasmodia (Malaria)	116
	B. Babesia (Babesiosis)	119
	C. Trypanosomes	120
	1. African Trypanosomiasis	120
	2. Experimental Infections with Nonhuman Trypanosomes	121
	3. *Trypanosoma cruzi* (Chagas' Disease)	122
	D. Leishmania	123
	E. *Entamoeba histolytica*	124
	F. *Toxoplasma gondii*	124
II.	Helminths (Metazoa)	125
	A. Cestodes (Tapeworms)	126
	1. Taenia	126
	2. Hymenolepis	126
	3. Echinococcus	126
	B. Trematodes (Flukes)	127
	C. Nematodes (Round Worms)	132
References		133

Chapter 7
Complement and Infectious Diseases 141

I.	Complement as a Normal Host Defense System	141
II.	Complement as a Mediator of Disease	142
	A. Sepsis	142
	B. Immune-Complex Disease Mediated by Complement	142
	C. Other Mechanisms of Complement-Mediated Inflammation	144
III.	Comments	145
References		145
Index		149

Chapter 1

THE COMPLEMENT SYSTEM

I. THE COMPLEMENT SYSTEM

The complement system is a group of plasma proteins that interact to produce a variety of physiological and pathological effects. The effect(s) produced in a given circumstance depends upon the substance activating the complement system, the step at which the system is activated, and the point at which activation terminates.

A general principle is that the various proteins or components of the system circulate in inactive form. One component upon activation may then activate the next component in the sequence or "cascade". Activation may involve internal rearrangement or cleavage of the protein so as to expose functional, usually enzymatic, sites, or aggregation of components may produce an effective complex. Natural circulating inhibitors compete with substrate proteins for sites on components and overall function represents the balance among activators, substrates, and inhibitors.

Complement proteins are produced in vivo and in vitro by a variety of tissues and cell types, including hepatocytes,[1] liver tissue,[2,3] reticuloendothelial cells,[4-7] and small intestinal epithelial cells.[8,9] At least for the third component, though multiple cell types can be demonstrated in vitro to produce the protein, the hepatocyte appears to be the primary in vivo site of synthesis.[10] Catabolism involves constant and relatively rapid decay of the proteins, perhaps 1 to 3% of the plasma pool per hour.[11]

A. Terminology

The complement literature bristles with complex nomenclature, which serves as a major barrier to wider understanding. Proteins of the system are listed in Table 1. Certain principles apply to general nomenclature. Complement is indicated by the capital letter "C" (the older use of "C'" has been discarded). Components of the classical pathway and terminal sequence (see below) are indicated by "C" for complement followed by a number representing the order in which the component was discovered; thus, C1 was discovered first, then C2, etc. Regrettably, these components do not react in the precise order in which they were discovered. Terminology of the alternative pathway components (see below) remains unsettled, though custom and preliminary agreement have strongly favored use of the properdin system terminology.[12] By this usage, proteins are called "factors" and distinguished by capital letters (e.g., factor B, factor D); properdin itself is called only "P".

Activated components may be indicated by a bar above the letter (e.g., $\overline{C1}$). Alternatively, cleaved subunits of components may be indicated by lowercase letters (a, b, c, d, etc.); some of these subunits are active (e.g., C3b, C5a), others inactive degradation products (e.g., C3d). Inhibitors or inactivators may be designated by functional (e.g., $\overline{C1}$ INH, C3b INA) or biochemical (β1H) names. Figure 1 illustrates these principles using C3 as an example.

B. C3

The third component of complement, C3, is present in serum in highest concentration, 1200 µg/dℓ.[11] Sometimes referred to as β1C in recognition of its migration

Table 1
COMPLEMENT PROTEINS

Component	Synonyms	Breakdown products
Classical pathway		
C1		
C1q		
C1r		
C1s	C1 esterase	
C4	β1E	C4a, C4b
C2		C2a, C2b
Alternative pathway		
Factor B	C3 proactivator (C3PA)	Ba, Bb
	Glycine-rich β-glycoprotein (GBG)	
Factor D	C3 proactivator convertase	
P (properdin)		
C3	β1C	C3a, C3b (β1A), C3c, C3d, C3e
Terminal sequence		
C5	β1F	C5a, C5b
C6		
C7		
C8		
C9		
Inhibitors		
C1̄ INH	C1 esterase inhibitor	
β1H		
C3b INA		

on serum protein electrophoresis, this molecule occupies a central position in the complement system, both functionally and biochemically (Figure 2).

Activation of C3 (about 180,000 dalton) occurs upon cleavage into the two subunits C3a (about 10,000 dalton) and C3b (about 165,000 dalton) (Figure 1). Enzymes capable of cleaving C3 (C3 convertases) may be generated through the classical or the alternative pathway. The antigen upon which the convertase is generated becomes the site of deposition of nascent C3b particles; the "acceptor" sites do not appear to be specific "receptors" for C3b but rather nonspecific but stable sites to which C3b is bound by hydrophobic and by either ester or imidoester bonds.[13]

Many molecules of C3 may be activated by a single convertase.[11] Some of these molecules will remain in fluid phase, where they are rapidly inactivated. Inactivation involves first the binding to C3b of β1H globulin, a normal serum protein, following which C3b inactivator (C3b INA) cleaves and degrades C3b.[14,15] Law et al.[15] suggested that C3b INA cleaves C3b to C3b′, a molecule that is antigenically intact but that has lost much of its functional activity. In subsequent degradation, proteolytic enzymes such as trypsin or plasmin cleave C3b′ to C3c and C3d.

Molecules of C3b that become bound to acceptor sites on activating antigens are presumably partly protected from the effects of β1H and C3b INA, so that they may mediate the biological effects of C3b. Activity of C3b depends therefore on the balance among the number of C3 convertase sites, continued activity of the C3 convertases (that have their own inhibitors), numbers of C3b molecules bound to "acceptor" sites, and relative protection of C3b from interaction with β1H. The amplification built into the complement system at several steps, including C3 activation, presumably represents an important part of this balance.

One important function of C3b is to initiate the terminal sequence of the comple-

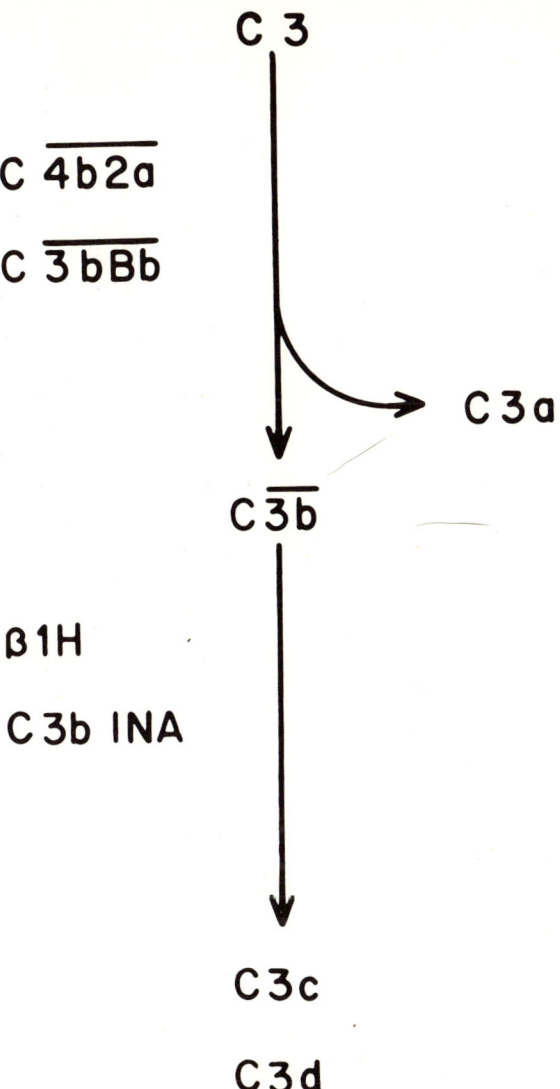

FIGURE 1. Principles of complement nomenclature as illustrated by activation-inactivation of C3. The native molecule (C3) can be cleaved by either the classical pathway or the alternative pathway C3 convertase (C4b2a or C3bBb, respectively). The smaller fragment (C3a) is soluble and rapidly inactivated. The larger fragment (C3b) may become bound to membranes or immune complexes; the fact that it is an active molecule can be indicated by the bar (C$\overline{3b}$). Natural inhibitors (β1H, C3b INA) inactivate the molecule, in this case by cleaving C3b into the inactive subunits C3c and C3d.

ment system (see below). More important is that C3b may provide a ligand between the antigen to which it is bound and certain cells and tissues with specific receptors for C3b.[16] These receptor-bearing cells include polymorphonuclear leukocytes,[17] monocytes and macrophages,[18,19] B lymphocytes,[20] human erythrocytes,[21] platelets from some species other than man,[17] and renal glomeruli.[22] The functions of these receptors and the C3b ligand will be discussed in subsequent sections and chapters.

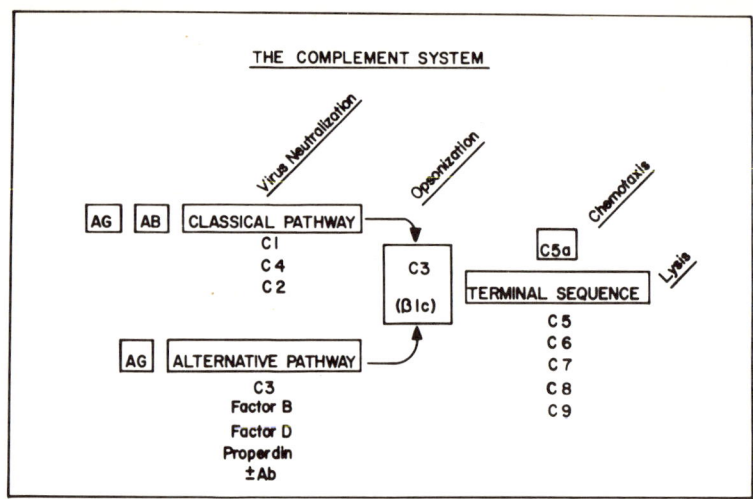

FIGURE 2. Schematic diagram of the complement system.

C. The Classical Pathway

Hemolysis of sheep erythrocytes, the standard model for the complement system, is based upon activation of the classical pathway. In this model, the Forssman antigen on the erythrocyte surface reacts specifically with rabbit antibodies, which then arm C1, the first component of complement. C1 is a trimolecular complex bound by calcium ions; the three subunits are designated C1q, C1r, and C1s. C1q binds to exposed sites of the Fc fragments of immunoglobulins G or M; internal rearrangement of C1 results in expression of the C1s-associated esterase (C1 esterase or $\overline{C1}$). The C1 esterase can then cleave C4 into its subunits C4a, which remains in fluid phase, and C4b, which remains membrane-bound. The combination of C1s and C4b cleaves C2. The smaller fragment, C2b, is dissipated; the larger fragment, C2a, becomes bound to the C4b molecule to form the classical pathway C3 convertase, i.e., the enzymatic activity, generated through the classical pathway by antigen-antibody complexes, capable of cleaving C3.

For every immunoglobulin-C1q complex, there may be multiple membrane-bound C4b2a complexes, another example of the characteristic amplification of the complement system. The active $\overline{C1}$ (C1 esterase) can be inhibited by the C1 esterase inhibitor. Inhibitors of C4b have been described.[23] C2a is a highly unstable molecule, undergoing decay from the complex in a matter of minutes.[11] Thus, even within the province of the classical pathway, one can see the balancing principles of amplification and inhibition.

While no definite physiological or pathological role has been identified for either C4a or C2b, the smaller fragments released upon activation of C4 and C2, they may have kinin-like properties of vasodilation.[24] Otherwise, with the exception of potential roles in viral infections (see Chapter 5), the classical pathway proteins serve primarily intermediary functions through activation of C3 and the terminal sequence.

D. The Alternative Pathway

In 1954, Pillemer and colleagues[12] described a system in human serum that appeared different from the classical hemolytic pathway. The distinctive protein of this system was properdin, which could react with a crude preparation of yeast cell walls (zymosan) to activate C3, bypassing C142. Later, they suggested that bacterial cell walls, endotoxin, and neutral polysaccharides could also activate this system,[25] which had great potential importance to immunobiology. However, methods available then did not permit answers to stringent criticism of the data.[26,27] Then, in the late 1960s and early 1970s, several lines of investigation established that there must be an alternative to the classical pathway. Guinea pigs deficient in C4 were healthy and their serum supported complement activation by substances such as zymosan.[28,29] Whereas guinea pig antibodies of the γ2 type activated complement beginning with C1, γ1 antibodies appeared to bypass C142 and activate C3.[30] Endotoxin could deplete serum of C3 by a mechanism independent of C2.[31] Final confirmation came from the protein chemical studies of Götze and Müller-Eberhard.[32]

Since these landmark studies, the "alternative" or "properdin" system of complement activation has been reconstructed from highly purified components and indeed shown to be capable of functioning completely independent of antibodies and the classical pathway components C1, C4, and C2.[33,34] Proteins comprising this pathway include C3 (which serves as both activator and substrate), factor B, factor D, and properdin (P). The substrate for active factor D ($\overline{D}$) is factor B; for active factor B ($\overline{B}$ or Bb), C3. The complex of C3bBb is autocatalytic, providing positive feedback on the pathway. Known inhibitors include C3b inactivator (C3b INA) and β1H globulin.

The best current hypothesis of the biochemical function of the alternative pathway derives from investigations by Fearon and Austen.[33,35] They demonstrated that factor D was normally present in serum as a fully potent enzyme ($\overline{D}$). Combination of purified preparations of $\overline{D}$, B, C3, and P resulted in spontaneous activation of the whole system with cleavage of factor B into its subunits Ba (a fluid-phase protein of no known activity) and Bb, C3 into C3a and C3b, and formation of C3bBb complexes stabilized by P. Addition to the system of β1H, which binds competitively to C3b, blocked formation of C3bBb complexes and inhibited further activity. Addition of zymosan negated the effect of β1H and permitted alternative pathway function. Likewise, C3b bound to sheep erythrocytes (which do not activate the alternative pathway under usual conditions) was readily inactivated by β1H and C3b INA, whereas zymosan-bound C3b was resistant. Other alternative pathway activators, such as Gram-negative bacteria and rabbit erythrocytes, also protected C3b. The suggestion was that C3b generated spontaneously in vivo (and ordinarily rapidly inactivated by β1H and C3b INA) could find a "privileged site" on surfaces such as zymosan and thus participate in amplification of the system. The biochemical basis for the protection appeared to be enhanced affinity for nascent C3b of factor B relative to β1H.[36] Thus, whereas the relative affinity of β1H for C3b might ordinarily be greater, when C3b was bound to zymosan (for example) factor B would have the greater affinity for C3b. Observing that alternative pathway activators were deficient in sialic acid,[36,37] these investigators suggested that sialic acid either enhanced the affinity of β1H for C3b or decreased the affinity of factor B, in either case impeding activation of the alternative pathway. Treatment of sheep erythrocytes with sialidase to remove sialic acid produced cells able to activate the alternative pathway, the degree of activation correlating with the amount of sialic acid removed.

The role of antibody in alternative pathway activation is poorly understood but of immense theoretical importance. Fearon and Austen[33] clearly demonstrated, using purified components, that the pathway may function in the absence of antibody. However, immunoglobulins may enhance activation of the alternative pathway[38] and, in some systems, may be required. The activating site is in the F(ab)$_2$ region of the molecule.[39–41] Antibody classes shown to be capable of activating the alternative pathway include IgG,[30,39,40] IgA,[41,42] and IgE.[43] IgA and IgE must generally be altered (for example, by aggregation) before they are effective activators, though these alterations may have in vivo counterparts.[41]

The theoretical importance of any antibody dependence of the alternative pathway lies in consideration of the physiologic role of the pathway. Evidence exists that functional components of this pathway phylogenetically antedate immunoglobulin and the classical pathway.[44] That observation, plus the demonstration that activation may occur in the absence of antibody, suggests that the alternative pathway functions as a primary though perhaps primitive defense mechanism against a wide variety of cells and surfaces. While antibody might enhance this system, it would not be necessary. On the other hand, if specific antibody were required for *effective* function of the system, then any potential role of the alternative pathway in early host defenses (i.e., prior to the appearance of specific antibody) would seem minimal. Other possibilities are that immunoglobulins adhering nonspecifically or cross-reactive antibodies of low affinity might play critical roles in alternative pathway function.

E. Terminal Sequence

Active membrane-bound C3b, generated through either the classical or alternative pathway, cleaves C5 into the subunits C5a and C5b and thereby initiates the terminal sequence. The fluid-phase fragment C5a, of approximately 10,000 dalton mol wt, has two important activities.[45] First, it induces vasodilatation and increased vascular permeability. In this activity, C5a (as well as the similar but less potent C3a) can be called an "anaphylatoxin." C5a is also a chemoattractant for polymorphonuclear leukocytes and macrophages. These subjects will be discussed more fully below (Section III).

Remaining proteins of the terminal sequence accumulate around membrane-bound C5b to form a macromolecular structure which has been termed the "membrane attack mechanism". This complex consists of one molecule each of C5b, C6, C7, and C8. Addition of up to six molecules of C9 completes the process with production of transmembrane channels through which ions and water may pass freely. The natural consequence of the membrane incompetence is usually lysis of the cell on which the reaction has occurred.[46-48] Details of this reaction as it relates to lysis of Gram-negative bacteria will be considered in the section on serum bactericidal reactions.

II. BIOLOGICAL ROLES OF C3 RECEPTORS

A. Opsonization

As a general rule, microorganisms cannot be ingested and killed by phagocytes in the absence of some binding force.[49] This concept is readily demonstrated in commonly employed in vitro phagocytosis systems (Figure 3). If living bacteria and polymorphonuclear leukocytes are incubated in a test tube, in roughly equal numbers, for 1 or 2 hr, rotating for adequate contact, bacteria will not be appreciably ingested and killed by the leukocytes. In such a system, it is true, bacteria of some strains may be killed, but these bacteria are almost always of low pathogenicity. The rule is that phagocytes do not by themselves ingest pathogenic microorganisms.

Some have suggested that the surface properties, especially surface charge, of these two particles, microorganism and phagocyte, may be critical in their mutual unattractiveness.[50,51] Reduction of surface changes may be a necessary first step in phagocytosis. Another possibility is that contact angles established between the two particles may determine subsequent events.[52] The low contact angles observed with

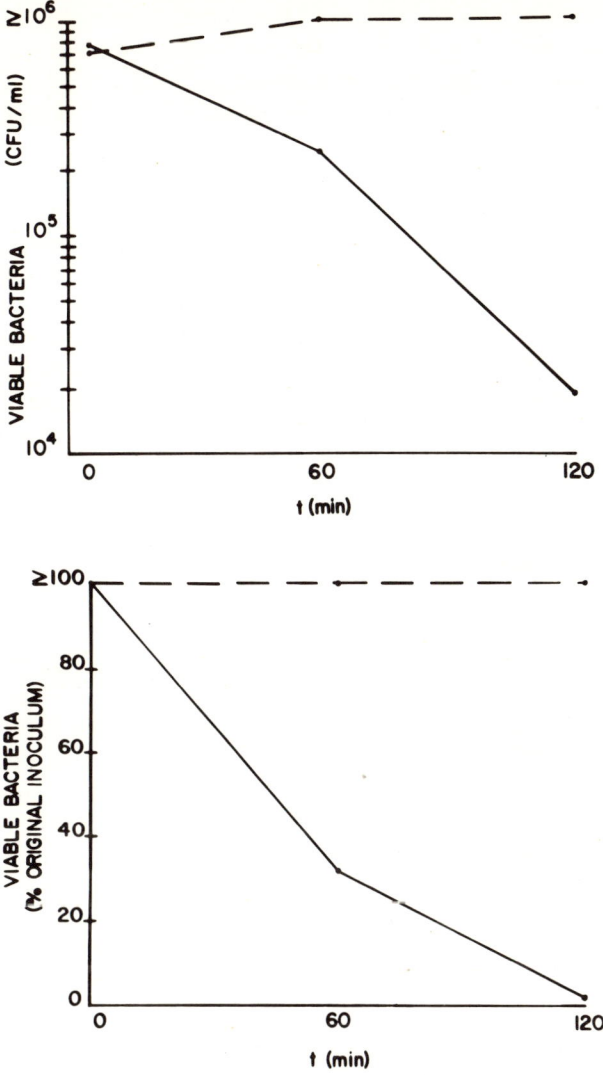

FIGURE 3. Phagocytosis of *Streptococcus pneumoniae* by human polymorphonuclear leukocytes. Pneumococci (approximately 2×10^7/mℓ; 0.1 mℓ) and neutrophils (5×10^6/mℓ; 0.5 mℓ) were mixed in polypropylene tubes with sufficient buffer to bring the total volume to 1 mℓ. Some tubes also contained 10% (final concentration) variously treated serum. Tubes were rotated at 12 r/min at 37°C. At 0, 60, and 120 min, aliquots were withdrawn, diluted in distilled water to lyse neutrophils and release intracellular but still viable bacteria, and cultured quantitatively. Results are expressed as either the absolute number of viable bacteria [colony-forming units (CFU)/mℓ] at each time (A) or the percentage of the original inoculum (the count at time 0) at each time (B). In the presence of 10% normal human serum (——), more than 6×10^6 bacteria or 97% of the original inoculum were killed. In the absence of serum or leukocytes, in the presence of heat-inactivated (56°C, 30 min) serum, or in the presence of serum absorbed at 0°C with pneumococci to remove antibody (all these conditions indicated by the interrupted line, ---), bacteria were not killed and usually proliferated.

some pathogenic, noningestible bacteria may relate to hydrophobicity induced by the capsules.

Methods to overcome repelling forces and induce closer approximation (for example, packing by centrifugation or addition of dextran) may be sufficient to induce ingestion of particles not generally considered ingestible.[53] Wood and colleagues[54,55] demonstrated that polymorphonuclear leukocytes seemed to ingest pneumococci if there was a surface against which the leukocytes could trap the bacteria. However, these observations have not been reproducible in some laboratories,[56,57] and the nonspecific closeness induced by centrifugation or dextran has no clear physiological counterpart.

In the experimental system outlined above, in which bacteria and phagocytes tumble in relatively dilute concentrations, ingestion and killing of 95% or more of the bacteria may readily be induced by introduction of serum or certain serum substances (Figure 3). Serum factors capable of this remarkable effect are called "opsonins," a term derived by Wright and Douglas in 1903:[58]

> We have here conclusive proof that the blood fluids modify the bacteria in a manner which renders them a ready prey to the phagocytes.
> We may speak of this as an "opsonic" effect (opsono — *I cater for; I prepare victuals for*), and we may employ the term "opsonins" to designate the elements in the blood fluids which produce this effect.

Opsonins provide ligands between the particle to be ingested and the ingesting phagocyte. The binding to the phagocyte may generally be considered to involve specific receptors for the opsonin. Miler[59] has emphasized that opsonization is an example of the much larger concept of cell-cell, cell-surface, and cell-particle interactions. For example, embryogenesis and its organization of cells depend on surface proteins, endogenous or exogenous to the cell, and their neutralization of forces such as negative charge.

Human cells with phagocytic capacity include polymorphonuclear leukocytes (neutrophils) and macrophages. The latter may be fixed in various tissues of the reticuloendothelial system (Küppfer cells of the liver, splenic macrophages, alveolar macrophages) or they may wander. Peripheral blood monocytes are macrophages in transit. Polymorphonuclear leukocytes are capable of more rapid ingestion and killing than are macrophages, but opsonic receptors and requirements appear similar for both cell types.[60] Macrophages and neutrophils have complementary roles in host defenses, and in discussing their behavior as phagocytes, we need not address their relative importance.

The serum opsonic activity described by Wright and Douglas[58] was heat labile, i.e., was lost upon heating of serum at 56°C for 30 min. This lability, while generally assumed to implicate complement, cannot alone be taken as proof that complement was the opsonin described by Wright and Douglas. After publication of their observations in 1903 and 1904,[58,61] considerable debate ensued as to the identity of the serum opsonins. A complete review of this debate is unnecessary. It is sufficient to point out that antibody was soon demonstrated to be a major opsonin, especially of immune serum, and subsequent conflict has largely revolved around the role (if any) of complement as an opsonin. Ward and Enders,[62] in 1933, suggested that antibodies were the major pneumococcal opsonins, even in unheated normal serum. Ecker and Lopez-Castro,[49] in 1943, concluded that complement played an important role in opsonization of *Staphylococcus albus* and *Staphylococcus aureus*. But, as recently as 1965, a major review could not identify any clearly defined role for the complement system in opsonization.[63]

Before considering the voluminous literature on this subject since 1965, one should first review in some detail the methodology of opsonic studies. Most early studies of phagocytosis used morphological methods. Particles to be ingested, a source of opsonins, and phagocytes were incubated, usually at 37°C, usually with rotation or frequent agitation, for some predetermined time, usually 30 min to 2 hr. A drop of the mixture was then placed on a glass slide, fixed, stained, and examined under light microscopy. Either the percentage of phagocytes with associated particles or the average number of particles per phagocyte was counted and the results expressed as a phagocytic index. Some more recent investigations have also used essentially the same methods. For large particles, such as erythrocytes, these methods appear to be reliable, especially when electron microscopy supplements the light microscopy.[64] With small particles such as bacteria, accurate morphological quantitation of phagocytosis, especially by light microscopy, is extremely difficult and perhaps impossible.[65] In the first place, bacteria that are adherent to the microscope slide may appear adherent to the phagocyte. Second, adherent bacteria cannot be distinguished from ingested bacteria. Third, bacteria cannot be reliably distinguished from neutrophil granules. Bacterial phagocytosis data obtained using only morphological techniques must be interpreted cautiously.

Bacterial killing by polymorphonuclear leukocytes or macrophages in vitro has generally become accepted as the most reliable assay for opsonization. Some variation of the method described earlier in this chapter is used.[66] Viable bacteria, phagocytes, and an opsonin source are incubated on a rotator at 37°C. Bacteria and opsonins are sometimes incubated together, after which the bacteria are washed and then incubated with phagocytes. This preopsonization allows for separation of the events of opsonization and phagocytosis, but phagocytosis is generally not so efficient in such a system. Aliquots of the mixture are removed serially for quantitative cultures. Results may be expressed directly as log numbers of viable bacteria or indirectly as percentage of the original inoculum. In the absence of opsonins or phagocytes, bacteria multiply. In the presence of adequate opsonins, bacteria are killed; generally, more than 90% of the bacteria can be killed in 2 hr. Because dilutions for the quantitative cultures are usually made in water, the phagocytes lyse osmotically and cultures reflect viable intracellular as well as extracellular bacteria. By differential sedimentation of the heavier phagocytes, one may quantitate both cell-associated and noncell-associated bacteria. A limitation of this methodology is that one cannot generally distinguish cell-associated from intracellular organisms. Thus, the attachment and ingestion phases of phagocytosis cannot be dissected. Furthermore, since organism death is the end point, one cannot separate intracellular killing from ingestion.

Interpretation of the literature is hampered by the fact that multiple variations of this phagocytic assay have been used. Each variation may alter the results. The phagocyte to bacterium ratio may vary considerably, from 10:1 to 1:1 to 1:10. The dilution of opsonin is important; most investigators have used a final concentration of serum (or other opsonin source) of 10 or 20%, but percentages have varied from 0.5 to 40. Duration of incubation varies. No buffer is standard. Sources and preparation of phagocytes vary. Perhaps most importantly, the susceptibility of bacteria to opsonization and phagocytosis may be affected by such things as phase of bacterial growth at harvest, the culture medium, and the amount of washing prior to incubation with phagocytes.

A similar method for studying phagocytosis uses bacteria that have been radiolabeled. After incubation as described above, phagocytes and unattached bacteria are separated by differential centrifugation, and cell-associated radioactivity is determined.[67] This method is technically easier but offers no conceptual advantage.

More recently, several ingenious phagocytosis assays have been developed to exploit the metabolic events that take place inside phagocytes after attachment or ingestion of particles. These methods include measurement of nitroblue tetrazolium reduction by phagocytes,[68] conversion of ^{14}C-glucose (labeled in the "1" position) to $^{14}CO_2$ during activation of the hexose monophosphate shunt,[69] generation of superoxide during[70] and chemiluminescence originating from relaxation of carbonyl groups following[71] superoxide-mediated bacterial killing. All of these metabolic events are part of the microbicidal activity of the phagocyte and, thus, like the killing assays described above, are only indirect measures of opsonization and ingestion.[72] Furthermore, since these changes may in some cases be triggered by membrane surface events such as receptor stimulation, these methods are also unreliable for distinguishing attachment from ingestion.

In vivo opsonization and phagocytosis have been studied by a number of investigators.[73,74] The methods generally involve intravascular or intraperitoneal injection of organisms and subsequent sampling of blood or tissues to determine clearance or localization. Organisms may be viable or radiolabeled, preopsonized or not; experimental animals may be depleted of certain opsonins or immunized to enhance opsonins. These techniques have also been adapted to in vitro experiments using isolated perfused organs.[75]

This discussion of variation and limitation of opsonophagocytic assays is a necessary prelude to a review of the data regarding complement as an opsonin. The most complete and definitively studied system is the phagocytosis of sheep erythrocytes. Gigli and Nelson[76] demonstrated minimal phagocytosis by guinea pig peritoneal leukocytes of sheep erythrocytes sensitized with approximately 3000 IgG molecules per cell. Deposition of complement components C1, C4, C2, and C3 markedly enhanced phagocytosis. Enhancement did not occur until C3 was added but required the earlier components; quantitative data suggested that as few as ten C3-containing sites on the erythrocyte surface could suffice. Subsequent investigators have suggested however that complement only mediates attachment and that antibody is required for ingestion. Mantovani et al.[77] confirmed that C3 markedly enhanced ingestion of IgG-sensitized erythrocytes by mouse peritoneal macrophages; however, they could not demonstrate ingestion of IgM-sensitized erythrocytes even in the presence of complement. Messner and Jelinek[78] confirmed the complement-independent ingestion of IgG-sensitized erythrocytes by human monocytes and polymorphonuclear leukocytes. Mouse neutrophils also ingested only IgG-sensitized erythrocytes though complement enhanced the reaction.[79] Ehlenberger and Nussenzweig[53] performed definitive experiments, which established that the Fc receptor of human neutrophils and monocytes mediated erythrocyte ingestion but was relatively inefficient at binding erythrocytes, whereas C3 markedly enhanced binding but could not mediate ingestion. For phagocytosis of some particles, they suggested, complement might be a necessary though not sufficient reactant.

The effect of complement in enhancing phagocytosis of erythrocytes could be mimicked by measures that increased erythrocyte-phagocyte contact: centrifugation of cells so that they were tightly packed, addition of high-molecular-weight dextran or protamine to the reaction mixture, or neuraminidase treatment of erythrocytes. Neuraminidase treatment or the presence of polycations may, among other things, counteract the normally repulsive electronegative charges on cells. Thus, Ehlenberger and Nussenzweig[53] suggested that the primary opsonic effect of complement is to overcome electronegative charges and lead to closer approximation of cells.

Griffin and colleagues[80,81] demonstrated ingestion by *activated* macrophages of erythrocytes sensitized with IgM and complement. They concluded that, unlike neutrophils or resting macrophages, activated macrophages could ingest particles upon stimulation of the complement receptor. However, Ehlenberger and Nussenzweig[53] subsequently demonstrated that, in the complete absence of complement, activated macrophages could be induced to ingest erythrocytes by nonspecific approximation. They issued a challenge, that "Claims that binding of C3 to a particle triggers the phagocytic act directly require the demonstration that the particle is not intrinsically ingestible; that is, that interiorization cannot be achieved by simple approximation of particle and phagocyte by physical means."

Griffin and colleagues[64,81,82] have looked more closely at the actual mechanisms of attachment and ingestion. They showed that particles such as erythrocytes attached to macrophages at sites other than the Fc receptor (e.g., at the C3 receptor) were not ingested during ingestion of latex particles (which do not require opsonins for ingestion) or of IgG-coated pneumococci. They concluded that the two receptors could function independently and that ingestion is a very select and localized process. In a further set of experiments, they mixed opsonized erythrocytes with macrophages at 4°C, during which time the erythrocytes could attach though the temperature was too low for ingestion. After washing the macrophages to remove unattached erythrocytes, they either blocked unoccupied Fc receptors with anti-macrophage IgG or inactivated unoccupied complement receptors with trypsin. Then, the macrophages were warmed to 37°C and ingestion was observed. IgG-sensitized erythrocytes were not ingested by macrophages with blocked Fc receptors. Likewise, erythrocytes sensitized with IgM and complement were not ingested by trypsin-treated activated macrophages. They concluded that attachment of a particle at a single site was not sufficient for ingestion; rather, a sequential attachment at adjacent receptors was required. They likened this process to closure of a zipper.[64] It is possible that the Fc and C3 receptors function synergistically in this zippering phenomenon.[53]

In experiments with rat Küpffer cells, Munthe-Kaas et al.[83] observed morphologically different phagocytic mechanisms for IgG-sensitized and for IgM-complement-sensitized sheep erythrocytes. Under the electron microscope, cytoplasmic projections rose rapidly to surround IgG-sensitized erythrocytes, in a manner consistent with the "zipper" hypothesis of Griffin et al.[64] This phagocytosis was inhibitable by cytochalasin B, a reagent that inhibits microfilaments. Internalization of complement-coated erythrocytes appeared to involve another mechanism not inhibited by cytochalasin B. The erythrocytes sank directly into the cytoplasm of the Küpffer cells; this process was slower than the phagocytosis of IgG-sensitized erythrocytes. Since fetal calf serum was required for Küpffer cell function, antibody may have been present in all reaction mixtures. Thus, the authors could not claim to have studied antibody-independent, complement-mediated phagocytosis. Nevertheless, they suggested that antibody-mediated phagocytosis appeared to be mechanistically different from that mediated by complement (with or without antibody). Kaplan made similar observations using activated mouse macrophages.[84]

A critical question is whether Fc receptors can play any important role in physiological circumstances. Mantovani et al.[77] demonstrated that normal IgG competes with antigen-bound IgG for the receptors in vitro and in vivo. Furthermore, Messner and Jelinek[78] found that the opsonic activity of erythrocyte-bound IgG could be inhibited by C1 or C1q in concentrations well below those found in serum. Thus, in physiological situations, IgG may not be able to function directly as an opsonin. In this regard, it might be remembered that most in vitro phagocytic assays utilize highly diluted serum (5, 10, or 20% serum concentration); such assays may be far from physiological. The primary beneficial effect of immunization may be altering the molar ratios or relative binding efficiencies in favor of antigen-bound IgG so that antibodies may function directly as opsonins. Alternatively, the major phagocytic function of complement may be to approximate particle-bound IgG and Fc receptors so that free IgG can no longer compete effectively for Fc receptors.

When one turns from phagocytosis of erythrocytes to phagocytosis of other particles, he encounters considerably more diversity of experimental results. There has been a tendency to extrapolate results obtained with sheep erythrocytes; such an extrapolation may not be warranted. There is no *a priori* reason to assume that rules applying to large particles such as erythrocytes must also apply to other, and especially, much smaller particles.

Considerable evidence suggests that bacteria may be ingested and killed by phagocytes in the absence of immunoglobulin, though the experiments have not generally been conducted with controls suggested by Ehlenberger and Nussenzweig[53] or under circumstances in which absolutely all immunoglobulin could be proven to be absent. Sterzl[85] used serum from colostrum-free piglets raised in a germ-free environment; these animals had no demonstrable antibody. In the perfused rat liver, this serum could provide an opsonic source for clearance of *Escherichia coli*. Midvedt and Trippestad[86] obtained similar results in an in vitro assay. Using human serum from agammaglobulinemic patients, Williams and Quie[87] demonstrated intact opsonic activity for a strain of *S. aureus*, several strains of *S. epidermidis*, viridans streptococci, and *Serratia marcescens*. *Staphylococcus aureus* 502A and *Streptococcus faecalis* were opsonized, though not so well as in normal serum. Several other organisms (*E. coli*, *Pseudomonas aeruginosa*, and *Bacillus proteus*) were not opsonized. The opsonic activity was heat-labile and absorbable by immune complexes. In similar studies, Jasin[88] depleted hypogammaglobulinemic serum both of C1q by immunoprecipitation and of remaining immunoglobulin on a sepharose-rabit-antihuman immunoglobulin column. This serum could still opsonize strains of *E. coli* and *Staphylococcus aureus*. Guckian and colleagues[89] absorbed serum with several microorganisms to remove antibodies; they could not demonstrate diminished opsonic activity as a result of these absorptions. Peterson et al.[60] concluded that opsonic requirements of bacteria and other microorganisms are extremely variable. Some organisms require antibody for phagocytosis, others require both antibody and complement, still others appear to require only complement. In the last case, opsonization appears to proceed through the alternative pathway. Whether actual ingestion is triggered by complement receptor stimulation or by some other, antibody-independent mechanism has not been established.

Variables that have received relatively little attention but may be of profound importance are those of rate and efficiency. Complement activation proceeds more rapidly through the classical than through the alternative pathway.[90,91] Clark and Klebanoff[91] compared opsonic function of the two pathways in C4-deficient serum. At low serum concentrations (e.g., 3%), little opsonization of zymosan for polymorphonuclear leukocytes could be demonstrated; whereas, at 10% concentrations, C4-deficient and normal serum were equivalent. Even at 10%, however, opsonization took more time in C4-deficient serum. Furthermore, in systems where classical pathway activity was blocked, such as in C4-deficient guinea pig serum or human serum chelated with ethyleneglycoltetraacetic acid (EGTA), the rate of activation of the alternatve pathway was enhanced by the presence of immunoglobulins.[38,40] Thus, standard methods, which have relied on measurements of maximum opsonization (usually at 60 min), may exaggerate the importance of the alternative pathway and minimize any role for immunoglobulin. The natural history of patients with agammaglobulinemia implies that the alternative pathway alone cannot substitute for an intact humoral immune system.[92,93]

In summary, optimum opsonization occurs when specific immunoglobulin and complement, probably both pathways, can act in concert. For some substances, no other conditions will mediate phagocytosis. Opsonization of other substances may be accomplished through the alternative pathway, with or without contributions of immunoglobulins (perhaps cross-reactive) and other serum factors. Though this alternative pathway opsonization may not be optimal, especially in kinetic terms, it may be a critical factor in primary host defenses. Subsequent chapters will evaluate how virulence of microorganisms may relate to the ways in which they interact with the complement pathways.

B. Role of Complement in Intraphagocytic Killing of Microbes

No role for complement and other opsonins, beyond mediating attachment or ingestion by phagocytes, has been established. However, several investigators have asked whether opsonins might somehow be involved in the intracellular events that follow ingestion. The vacuole, formed by invaginating phagocyte membrane and containing the opsonized particle, fuses with the enzyme-laden lysosome, which releases microbicidal and digestive enzymes.[94] One can imagine that opsonins might facilitate these membrane interactions or that complement might enhance microbicidal events, much in the way the terminal complement sequence and lysozyme act synergistically in bacterial lysis.

Jenkin[95] suggested that opsonins (in these studies generally antibodies) facilitated the rate or degree of intracellular killing. In a study comparing clearance of serum-sensitive and serum-resistant *E. coli* in C5-deficient and C5-sufficient mice, Glynn and Medhurst[96] indicated a possible role for the terminal complement sequence in intracellular killing. The admitted problem of all who have addressed this question has been methodological. How can one study the role of opsonins intracellularly when variations in opsonins have such profound effects on earlier events in phagocytosis? The rate of intracellular killing must surely be affected by the rate of bacterial ingestion. One cannot remove opsonins from the reaction after ingestion; and indeed it is difficult to imagine how ingestion can be experimentally dissociated from intracellular killing since the latter begins very shortly after the former.

Using a complicated experimental system to manipulate immunoglobulins and the two complement pathways independently, Menzel et al.[97,98] suggested that complement enhanced intracellular killing of a serum-resistant *E. coli*. On the other hand, Van Snick and Masson[99] could demonstrate no effect of complement on intracellular digestion of immune complexes.

More recently Leijh and co-workers[100] have suggested that complement, activated through the alternative pathway, and to a lesser extent, immunoglobulin are necessary in the extracellular milieu for intracellular killing by human monocytes. Preopsonized bacteria (i.e., bacteria previously incubated in an adequate opsonic source) were mixed with monocytes briefly (3 min), during which time a majority of bacteria were ingested but essentially none were yet killed. Monocytes were rapidly separated from bacteria still in suspension; in the experiments with *S. aureus*, all extracellular but cell-associated bacteria were destroyed with lysostaphin. Then monocytes were suspended in various media and incubated further at 37°C. If the final suspending medium contained normal serum, intracellular killing proceeded effectively. Killing was impaired if complement and, more specifically, the alternative pathway were removed or blocked. Killing was also at least partially stimulated by immunoglobulin in the mixture. These authors concluded that extracellular complement and immunoglobulin were required for efficient intracellular killing of bacteria. Guckian and colleagues[101] have confirmed these observations regarding such a role for complement in intracellular killing of *S. pneumoniae* serotype 25 by polymorphonuclear leukocytes.

At present, the role of complement in intracellular killing must be considered still unsettled. However, tantalizing clues suggest that proof of such a role may be forthcoming.

C. Other C3 Receptors

Interaction of C3-bearing particles with complement receptors on polymorphonuclear leukocytes and macrophages is an important part of phagocytosis, as discussed above. Other human cells with receptors for C3 include B lymphocytes, erythrocytes, and renal glomerular cells. The functions of these receptors are speculative.

1. Erythrocytes

Nelson[21] demonstrated the phenomenon of immune adherence in 1953. *Treponema pallidum*, in the presence of syphilitic rabbit serum as a source of antibody and normal human blood as a source of complement, adhered to the human erythrocytes. Both antibody and complement were required. Adherence to polymorphonuclear leukocytes was also seen. No adherence to sheep or rabbit erythrocytes could be demonstrated. Similar results were obtained with *S. pneumoniae* serotype 1. He also found that phagocytosis of pneumococci was enhanced by the addition of human erythrocytes to the mixture of antibody, complement, and phagocytes. Subsequently, he demonstrated that the erythrocytes were not themselves ingested by phagocytes and that immune adherence occurred in vivo in primates.[102] These investigations suggested that immune adherence to erythrocytes may be important in bloodstream clearance of microorganisms.

On the other hand, van Loghem et al.[103] suggested that immune adherence led to shortened erythrocyte survival in vivo. Though their experiments were done with noninfectious immune complexes (ovalbumin-antiovalbumin), the same phenomenon might underlie some hemolytic anemias seen with infections.

The earlier demonstrations required antibody and the four then-known complement components.[102] However, May et al.[104] reproduced the phenomenon with C4-deficient guinea pig serum, indicating that the alternative pathway could mediate immune adherence.

Little attention has been paid to the erythrocyte complement receptor in recent years. A review in 1975[105] attributed no known significance to immune adherence, at least as regards the erythrocyte. The subject seems worthy of further study.

2. B Lymphocytes

The adherence of complement-coated sheep erythrocytes is a commonly used assay to identify that subset of lymphocytes known as bursal(B)-equivalent or bone marrow-derived lymphocytes.[20] B lymphocytes also have receptors for the Fc fragment of IgG, which apparently function independently of the complement receptors.[106] These B lymphocytes are the precursors of antibody-producing plasma cells. That they should have receptors for C3 has stimulated considerable discussion of a role for complement in the induction of the humoral immune response.

Pepys[107] has reviewed the hypotheses and the data presented in their favor and in opposition. Dukor and Hartmann[108] had suggested that thymus-independent antigens are also activators of the alternative pathway and that complement might play a necessary role in triggering B-cell differentiation and antibody response to such antigens. With thymus-dependent antigens, the T lymphocyte itself might release complement-cleaving enzymes. Though several subsequent studies have failed to show any necessity for C3 in antibody response,[109,110] Pepys[107] has suggested that complement may enhance the binding of immune complexes, macrophages, and B lymphocytes and thus facilitate a maximum antibody response. Complement is envisioned as playing an important but not necessary role in antibody response. This subject is obviously not settled and continues to be controversial.[111,112]

A different phenomenon in which complement may interact with lymphocytes is that of antibody-dependent cellular cytotoxicity (ADCC). The effector cells are those lymphocytes called null cells, lacking the usual B- or T-lymphocyte markers. They may however be a subset of B lymphocytes. Target cells can include foreign erythrocytes, tumor cells, and virus-infected cells. Killing requires mediation by antibody directed against the target cell and presumably reacting with Fc receptors on the lymphocyte surface.[113] Antibody-dependent cellular cytotoxicity can be inhibited in vitro by subphysiological concentrations of free IgG. This inhibition may imply that the phenomenon has no in vivo relevance; however, Scornik[114] demonstrated that the addition of complement to the IgG-coated target cell prevented the inhibitory effect of IgG. He suggested that complement produced more intimate contact of target and effector cell, which might favor receptor binding of antigen-bound over free IgG. Whatever the explanation, complement, though not sufficient for the reaction, appears to enhance antibody-dependent cellular cytotoxicity[115-117] and may be necessary in vivo.

3. Renal Glomeruli

Gelfand et al.[22] first described a receptor for complement in the human renal glomerulus. Others have confirmed the presence of the receptor and identified some of its characteristics.[118,119] It has been suggested that the receptor localizes circulating immune complexes to the glomerulus and thus may be pathophysiologically important in immunologically mediated glomerulonephritides.[120-123] What physiological role such a receptor might serve remains conjectural. Against the notion that the glomerular complement receptor is a primitive or vestigial clearance mechanism is the failure to find such receptors in laboratory animals studied thus far.[118,124]

D. Variability Among Complement Receptors

Though one speaks of *the* complement receptor, it should be pointed out that receptors of different cell types do not behave identically. Macrophages, polymorphonuclear leukocytes, B lymphocytes, renal glomeruli, and primate erythrocytes will all bind C3b-bearing particles. C4-bearing particles will also bind to these cells, apparently at the same receptor,[125] though Bokisch and Sobel[126] inferred the receptors to be distinct since Raji cells bound C3b-coated but not C4-coated erythrocytes. The site on the C3b molecule that reacts with the C3b receptor apparently lies in the C3c portion.[125]

Some cells, including macrophages and some B lymphocytes, also have receptors for further degradation products of C3.[125] These receptors, which cap independently of C3b receptors and thus are on different molecules,[125] now appear to be of at least two species, one reactive with the C3bi molecule (an inactive intermediate degradation product of C3b), the other with the C3d molecule (produced on cleavage of C3bi by proteolytic enzymes such as trypsin or plasmin).[127]

The C3b receptor, which also reacts with C4 and C3c, is the immune adherence receptor and the one to which these comments apply.[128] The C3d receptor is demonstrable in vitro and can mediate adherence, e.g., to macrophages.[53] However, the normal in vivo clearance of C3d-coated erythrocytes implies that the C3d receptor may serve no important physiological role.[129]

III. ANAPHYLATOXINS, CHEMOTAXIS, AND STIMULATION OF LEUKOCYTES

The Arthus reaction is a commonly studied model of immunologically mediated tissue damage. It is produced by intravenous injection of antibody and subcutaneous injection of the appropriate antigen or vice versa. Within a short time the ensuing antigen-antibody reaction in local vascular walls results in accumulation of neutrophils, destruction of the vessel walls (due at least in part to leukocytic enzymes), and extravasation of erythrocytes. Complement components can be demonstrated in damaged vessel walls. The gross picture is that of acute inflammation with edema, hemorrhage, and perhaps necrosis.[130] Ward and Cochrane[131] depleted rats and guinea pigs of complement prior to injection of antigen and antibody. Complement depletion essentially aborted the Arthus reaction; there was no vascular damage, complement could not be demonstrated in vessel walls, and neutrophils did not accumulate around vessels. In a variety of other clinical and experimental situations, including the nephritis of systemic lupus erythematosus and experimental serum sickness, this close relationship can be demonstrated among complement activation, leukocyte accumulation, and tissue injury. The principal effect of complement activation appears to be the recruitment of leukocytes, primarily polymorphonuclear leukocytes, which release lysosomal enzymes, usually during phagocytosis, and damage host tissues as well as foreign antigens.[132] The complement system may also augment the inflammatory events associated with delayed hypersensitivity.[133]

In in vitro systems, complement activation generates factors that are attractive to leukocytes; that is, in a concentration gradient of these factors, leukocytes will migrate toward greater concentrations. Such directed migration of leukocytes is called "chemotaxis." Suggested complement chemotactic factors include C3a and C5a, the low-molecular-weight soluble split products of C3 and C5 activation, and the soluble complex of C5, C6, and C7.[134] Snyderman et al.[135] presented evidence that the major chemoattractant produced upon in vitro activation of fresh serum was C5a. In further studies,[136] they related the in vitro observations to in vivo studies of guinea pig peritoneal exudates. A chemotactic factor isolated from the exudates appeared identical to C5a; C5-deficient mouse serum could not generate chemotaxins in vitro; and intraperitoneal injection of endotoxin did not induce leukocyte accumulation in C5-deficient mice. This study clearly established C5a as the major chemotactic factor of human serum. With the subsequent purification and amino acid sequencing of C5a and C3a, it has been suggested that only C5a manifests significant chemotactic activity.[137] The normal chemotactic activity in serum from C6-deficient patients argues against any important chemotactic role for the C567 complex.[138]

In normal serum, C5a is actually rapidly degraded by carboxypeptidase B (or similar enzymes) to a form called $C5a_{des\,Arg}$.[45,137] Thus, even though C5a is highly chemotactic in vitro, its in vivo importance is probably minimal. However, $C5a_{des\,Arg}$, itself not potent chemotactically, is quite active in the presence of nonactivated serum,[137] which presumably supplies a "helper factor".[139] This $C5a_{des\,Arg}$-helper factor complex appears to be the major complement-derived chemotactic factor.[139]

Chemotactic activity may also be generated through the alternative pathway[140] though somewhat more slowly.[91,141] Furthermore, chemotactically active C3a and C5a may be generated directly from the parent molecules by proteinases produced by polymorphonuclear leukocytes,[142,143] macrophages,[144] heart, lung, liver, and spleen cells,[145] and several bacterial species.[146] Finally, other complement components may participate indirectly in chemotaxis: C3b can stimulate B-lymphocyte production of a chemotactic lymphokine[147,148] and C1 esterase inhibitor may sensitize leukocytes to C5a, such that an increased number of leukocytes respond in chemotaxis assays.[149]

Chemotactic effects are generally conceived of as occurring at very small distances. More systemic effects of complement upon leukocyte mobility were described by Rother,[150] who suggested that complement activation generated a factor, not chemotactic, which in vitro mobilized leukocytes from guinea pig or rat femurs and in vivo elicited a polymorphonuclear leukocytosis. McCall et al.[151] described a biological activity, apparently due to a low-molecular-weight factor produced upon complement activation by intravascular injection of inulin or cobra venom factor, which induced an immediate neutropenia followed by eventual (2 to 4 hr later) neutrophilia. More recently, Ghebrehiwet and Müller-Eberhard[152] have presented evidence that the C3e fragment mobilizes leukocytes from perfused marrow and produces a leukocytosis in vivo. The relationship of C3e to previously described complement derived, leukocytosis-promoting factors and the physiological role of C3e remain speculative but highly intriguing.

Perhaps related, especially to the transient neutropenia described by McCall et al.,[151] are observations that interactions of complement products with leukocytes can lead to enhanced granulocyte adherence to endothelial cells. It is known that granulocytes adhere to endothelial cells and that adherent fractions can represent as much as half the circulating pool. When granulocytes are exposed to complement fragments, probably C5a, they acquire increased "stickiness" and adhere in greater numbers to endothelium. This phenomenon has been most clearly demonstrated with the acute neutropenia associated with filtration leukophoresis[153] and hemodialysis.[154] More recently, Weisdorf et al.[155] have suggested this phenomenon underlies many cases of adult respiratory distress syndrome. And Goldblum and colleagues[156] have presented evidence that leukocyte adherence to pulmonary endothelial cells can be induced in rabbits by pneumococcal infection or sepsis. Sacks et al.[157] demonstrated that C5a-induced granulocyte adherence could result in endothelial cell damage, mediated at least in part by superoxide production.

Complement chemotactic factors may also affect leukocytes already at an inflammatory site. Issekutz et al.[158] found that zymosan-activated serum, partially purified C5a, and chemotactic peptide subunits of C5a could enhance phagocytic and bactericidal activities of polymorphonuclear leukocytes. The mechanism appeared to be membrane stimulation of oxygen-dependent bactericidal systems.

C5a and C3a are also potent anaphylatoxins, that is, they produce vasodilatation, increased vascular permeability, and smooth muscle contraction.[45,139] Though C5a is more potent than C3a, the greater concentration of C3 in serum provides at least the potential for C3a to be an important anaphylatoxin in vivo.[45] These molecules produce their effects, to a large extent, by directly stimulating mast cells and basophils to release histamine,[45,139,159] which mediator induces the smooth muscle and vascular changes. The smooth muscle contraction may also represent a direct effect of C5a and C3a.[45] Unlike the chemotactic effects, anaphylatoxin activity is lost when the terminal arginine is cleaved. Thus, in vivo complement-derived anaphylatoxins probably have little systemic effect and produce a "potent but transient tissue response."[45]

IV. COMPLEMENT-COAGULATION INTERACTIONS

The obvious biochemical similarities between the proteolytic complement and coagulation cascades have stimulated many efforts to demonstrate physiological connections. Such connections remain tenuous. The subject has been reviewed by Brown,[160] who found no evidence for any physiologically important interaction. Certainly, patients with complement deficiencies have had no disorders of coagulation (see Chapter 2). It is however possible that pathological degrees of either complement or coagulation system activation may lead to some activation of the other system, though perhaps not directly.[160]

Platelets from several subprimate species, including rabbits, have C3b receptors that generally seem to behave like other complement receptors. A very complex interaction of *human* platelets, zymosan, and human serum has been described, which may have relevance to human infection.[161] The reaction involves the alternative pathway, C567, fibrinogen bound to the zymosan, and immunoglobulin bound to zymosan and to platelets at the Fc receptors. The fibrinogen and complement must apparently be in close proximity. Conclusions regarding this phenomenon and its clinical relevance are not possible at this time.

V. LYSIS

Much of the knowledge of cell lysis by complement comes from analyses of the lysis of sheep erythrocytes. Kinetics and stoichiometry of this process have been studied at great lengths. Though considerable controversy exists regarding the exact nature of the transmembrane channel or lesion induced by the C5-9 complex, one may outline the general characteristics with considerable certainty. Activation of the classical pathway of complement by complexes of sheep erythrocyte antigens and rabbit antibody to sheep erythrocytes leads to distribution of complement "sites" around the antigens. These complexes of antigen-antibody-C142 cleave C3 molecules, some of which attach to the erythrocyte surface (see section on "C3") and cleave C5, initiating the terminal sequence. The rabbit antibody may be either IgG or IgM; it has been estimated that approximately 800 molecules IgG per erythrocyte are required for subsequent complement-mediated lysis.[162] The C5-9 complex, inserted into the erythrocyte membrane, in some way impairs the membrane integrity and produces lysis.

This antibody-complement-mediated lysis requires an intact classical pathway as evidenced by its failure to proceed in C4-deficient guinea pig serum.[163] Other sensitized (i.e., antibody-coated) cells may also be lysed by this classical pathway-dependent system: guinea pig thymocytes, mouse thymocytes, and guinea pig leukemia cells. In fact, virtually all animal cells are sensitive to the lytic effects of antibody and complement, as are most microbes; only Gram-positive and some strains of Gram-negative bacteria are resistant.[164,165]

As mentioned earlier, rabbit erythrocytes may be lysed by the human alternative complement pathway.[166] This reaction, though greatly enhanced by IgG, can proceed in serum depleted of immunoglobulins by repeated absorptions.[167] Schreiber et al.[34,168] have reconstructed a cytolytic system from highly purified alternative pathway and terminal sequence proteins; their system, which is not dependent on immunoglobulins, can lyse erythrocytes and Gram-negative bacteria.

Thus, activation of at least one of the complement pathways can, under specific circumstances, produce a lytic site in most microbial and animal cells. Examples of this phenomenon as they relate to specific organisms will be discussed in subsequent chapters. But some discussion of more general aspects of serum-mediated microbial lysis, the "serum bactericidal reaction," seems appropriate here.

Using a strain of *Salmonella typhi*, Rother et al.[169] demonstrated that serum bactericidal activity took place in two steps, behaving in general identically to the hemolytic system. In the presence of excess complement, killing was dependent on the dose of antibody; with optimium antibody, it was dependent on the concentration of complement. Both IgG and IgM antibodies can mediate bactericidal activity, but IgM appeared to be the more efficient.[170] Lysis of *Vibrio cholera* in rabbit, dog, or sheep serum could not be mediated by IgA antibodies,[170] as might be expected. However, Sirotak et al.[171] produced γ1 antibodies in guinea pigs by serial injections of *E. coli*; these antibodies were bactericidal primarily through the alternative pathway. They suggested that human IgA antibodies, analogous to guinea pig γ1, might also be bactericidal in some circumstances.

The membrane damage produced by complement on Gram-negative bacteria, and probably on other microbes as well, appears to be identical to that produced on erythrocytes and other eukaryotic cells. Bladen et al.[172] observed similar lesions by electron microscopy. The lesion appears to involve displacement of membrane lipids.[173]

It should be emphasized that microbial cell walls and membranes are considerably more complex than erythrocyte membranes (see further discussion in Chapter 3). Bacterial lysis requires, in some systems at least, the additional action of serum lysozyme,[174-177] which destroys components of the cell wall, after which final destruction of the cell membrane can occur through the effect of hypotonicity, complement, or other enzymes. These other enzymes might include beta-lysins of human platelets.[178,179]

Davis and colleagues[165] have stressed the importance of distinguishing bacterial killing from bacterial lysis. When serum and bacteria are incubated together, decreases in numbers of viable organisms can be seen within 10 min; this killing is mediated by antibody and complement. Lysozyme primarily mediates the subsequent appearance of spheroplasts, the round, fragile, cell-wall-deficient forms, which, by 30 min, are the predominant morphological forms. Thereafter, spheroplasts are transformed into bacterial ghosts, presumably due to total cell membrane destruction; concomitantly, the optical density of the bacterial suspension diminishes. Thus, importantly different conclusions as to the roles of various serum components will be reached depending on whether one studies true bacterial death or morphological changes of bacterial destruction.

Among Gram-negative bacteria, there is variability in serum sensitivity, i.e., in the susceptibility of the organisms to the bactericidal effect of serum. One must recognize that part of the variability may relate to changes within a given organism with time. Thus, resting bacteria of an otherwise sensitive strain are relatively resistant to the lytic activity of human serum; whereas organisms in the growth phase are rapidly killed. Furthermore, temperature requirements for growth and subsequent serum sensitivity vary.[180] And it should be noted that the literature is not highly consistent in other methodological matters, including concentration of serum in the bactericidal system, duration of exposure of serum and bacteria, and growth media for the bacteria. Nevertheless, as might be expected, variability in serum sensitivity among bacteria is largely related to the structure and composition of the cell wall. Procedures that strip cell wall components (e.g., prolonged incubation in chelators) can convert serum-resistant to serum-sensitive organisms; this finding may also apply to Gram-positive bacteria.[164,181] At the molecular level, it has been suggested that the cell wall, and especially the lipopolysaccharide portion, may mediate serum resistance primarily by blocking effective attachment of complement components, primarily C5, to the bacterial surface.[181,182]

The importance of variability in bacterial susceptibility to complement lytic activity will be discussed in subsequent chapters. In addition, chapters on fungi, viruses, and protozoa will address the importance of complement lysis in regard to other microbes.

VI. COMPLEMENT METHODOLOGY

The study of microbial interactions with complement has involved the use of multiple strategies and methodologies, the more widely used will be briefly reviewed at this point. Other experimental techniques will be considered in context of specific investigations.

A. Complement Assays
1. Functional Assays

The oldest and still most widely used functional complement assays are based on the lysis of sensitized erythrocytes. The mathematical foundations and methodology have been detailed by Mayer[183] and Rapp and Borsos.[184] Essentially, erythrocytes (usually sheep) are suspended to a standard concentration and "sensitized" with excess Forssman antiserum, prepared usually in rabbits by repeated injections of sheep erythrocytes. Dilutions of serum, or some other source of complement, are added to the sensitized erythrocytes and the mixture incubated at 37°C for 1 hr. The degree of lysis is calculated from spectrophotometric determinations of the amount of hemoglobin released from erythrocytes.

The standard functional hemolytic assay (CH50) requires all the components of the classical pathway (C142), C3, and the terminal components (C5-9). By supplying purified C1, C4, and C2 in excess, one can adapt this assay to measure only C3 and the terminal components ("C3-9"). And addition of all but a single component in excess allows functional assay of that component, since lysis will then be dependent on the concentration of the limiting component in dilutions of a clinical or experimental sample.[184]

Demonstration of lysis of rabbit erythrocytes in human serum[166] has also permitted functional hemolytic assays of the alternative pathway.[167,185] The most sensitive of these measure the kinetics of lysis as well as total lysis.[185]

One may also measure other functions of complement, opsonization, chemotaxigenesis, etc. But these methods have not been so well-developed quantitatively or so well-standardized as the hemolytic assays.

2. Immunochemical Assays

Using monospecific antisera, one may measure any complement component by standard methods of radial immunodiffusion and immunoelectrophoresis. The latter also allows some quantitative evaluation of the integrity of the component. For example, factor B as well as the active fragment Bb will react with antiserum to factor B; the different electrophoretic mobility of the two proteins allows one to distinguish them and to infer activation. Fluorescein-tagged antibodies to complement components can be used to identify the proteins deposited in tissue or on cells.

3. Metabolic Studies

The rate of disappearance from circulation of radiolabeled components following intravenous injection can be used to assess the rate of activation/decay.[186] Such studies, though difficult and expensive, have provided extraordinary insight into the dynamics of the complement system in various disease states. These techniques have allowed investigators to distinguish diminished synthesis of complement components from increased catabolism.

B. Strategies to Study the Role of Complement in Infectious Diseases

Perhaps the oldest method for studying involvement of complement in clinical or experimental infections is simply the direct measurement of serum complement levels. Because values vary greatly during various phases of any infection, serial determinations are more informative than isolated ones. Moreover, the increased synthesis of most components during acute inflammatory events (the "acute phase reaction") may obscure any consumption. Nevertheless, complement values during infections can provide useful data and can be a stimulus to more definitive investigations.

Some of these problems can be overcome by measuring products of individual components immunoelectrophoretically. Thus, immunoelectrophoretic conversion of factor B would provide additional evidence that a low complement level does indeed reflect activation. Metabolic studies of clearance of radiolabeled components, of course, provide even stronger data regarding activation or decreased synthesis. Deposition of complement components at inflammatory sites also indirectly suggests complement involvement. Complement values may be determined before and after microbes are incubated in vitro with serum. Consumption of complement, i.e., lower complement values after incubation with microbes, can then be attributed to activation by the microbe.

Simply measuring complement consumption by microbes does not provide data regarding the effect of complement activation. Thus, it is useful to measure also the generation of biologically active products (e.g., chemotactic factors) and the consequences to the microbe (e.g., opsonization, lysis, neutralization).

The amount of complement (usually C3) deposited on microbes following incubation with serum has more recently been studied by quantitative immunofluorescence.[187] The amount of fluorescence from a standard suspension of organisms can, by such studies, imply the amount of C3 deposited on the microbe.

C. Assessment of Complement Pathways

One of the major areas of interest in recent years has been the differential activation of the classical and alternative pathways by microbes. Some investigators have measured multiple complement components and inferred the pathway activated from the selective depressions of various components. For example, normal levels of C4 with low C3 would suggest alternative pathway activation. While useful, this approach is subject to considerable error; some antigens may be very efficient activators of the classical pathway, such that C3 is consumed by that pathway with relatively minimal consumption of C4 or C2. Moreover, activation of the classical pathway may, through nascent C3b, stimulate the alternative pathway with resultant consumption or conversion of alternative pathway proteins.

Selective blockade of a given pathway has proven useful for experimental evaluation. Use of congenitally deficient experimental animals has provided information of great interest regarding susceptibility to infection or manifestations of infection mediated by complement. Likewise, careful evaluation of naturally occurring infections in complement-deficient humans has provided insight into interactions of complement and microbes.

Complement may be inactivated in vivo by cobra venom factor, a highly potent activator of the alternative pathway.[188] Injections of suitable amounts can temporarily but completely deplete C3 and the terminal components.

In vitro, numerous methods have been used to deplete serum of individual components in order to assess their roles or the roles of the various pathways. Heating serum at 56°C for 30 min is a time-honored way of depleting both C2 and factor B; factor B but not C2 is at least partially labile at 50°C also. Various chemicals, such as hydrazine, may inactivate some components but are rarely used now. And, most specifically, immunoadsorbent columns may be used to remove components from serum if a monospecific antibody is available.

The classical pathway requires both calcium and magnesium, the alternative pathway only magnesium. This fact has been exploited by using differential cation chelation of serum with ethyleneglycoltetraacetic acid (EGTA).[189] This reagent has a much greater affinity for calcium than magnesium and thus selectively blocks the classical pathway. The alternative pathway is partially inhibited by this chelator since magnesium levels are subphysiologic; this problem can be overcome by use of the magnesium salt (MgEGTA).[166,190] Thus, measurements of complement-microbial interactions in MgEGTA-chelated serum provide data regarding the role of the alternative pathway.

Finally, some investigators have reconstructed functional complement pathways from purified components.[33,34,168] These studies have allowed definitive evaluation of potential in vivo functions of various components and pathways.

REFERENCES

1. **Strunk, R. C., Tashjian, A. H., Jr., and Colten, H. R.,** Complement biosynthesis *in vitro* by rat hepatoma cell strains, *J. Immunol.*, 114, 331, 1975.
2. **Hobart, M. J., Lachmann, P. J., and Calne, R. Y.,** C6: synthesis by the liver *in vivo*, *J. Exp. Med.*, 146, 629, 1977.
3. **Colten, H. R.,** Ontogeny of the human complement system: *in vitro* biosynthesis of individual complement components by fetal tissues, *J. Clin. Invest.*, 51, 725, 1972.
4. **Rubin, D. J., Borsos, T., Rapp, H. J., and Colten, H. R.,** Synthesis of the second component of guinea pig complement *in vitro*, *J. Immunol.*, 106, 295, 1971.
5. **Lai A Fat, R. F. M. and van Furth, R.,** *In vitro* synthesis of some complement components (C1q, C3 and C4) by lymphoid tissues and circulating leucocytes in man, *Immunology*, 28, 359, 1975.
6. **Littman, B. H. and Ruddy, S.,** Production of the second component of complement by human monocytes: stimulation by antigen-activated lymphocytes or lymphokines, *J. Exp. Med.*, 145, 1344, 1977.
7. **Barber, T. A. and Burkholder, P. M.,** Enumeration and ultrastructure of C4-producing free alveolar cells from guinea pig lung, *J. Immunol.*, 120, 716, 1978.
8. **Colten, H. R., Borsos, T., and Rapp, H. J.,** *In vitro* synthesis of the first component of complement by guinea pig small intestine, *Proc. Natl. Acad. Sci. U.S.A.*, 56, 1158, 1966.
9. **Colten, H. R., Gordon, J. M., Rapp, H. J., and Borsos, T.,** Synthesis of the first component of guinea pig complement by columnar epithelial cells of the small intestine, *J. Immunol.*, 100, 788, 1968.
10. **Alper, C. A., Johnson, A. M., Birtch, A. G., and Moore, F. D.,** Human C'3: evidence for the liver as the primary site of synthesis, *Science*, 163, 286, 1969.
11. **Kohler, P. F.,** Human complement system, in *Immunological Diseases*, 3rd ed., Samter, M., Ed., Little, Brown, Boston, 1978, Chap. 13.
12. **Pillemer, L., Blum, L., Lepow, I. H., Ross, O. A., Todd, E. W., and Wardlaw, A. C.,** The properdin system and immunity. I. Demonstration and isolation of a new serum protein, properdin, and its role in immune phenomena, *Science*, 120, 279, 1954.

13. **Law, S. K. and Levine, R. P.**, Interaction between the third complement protein and cell surface macromolecules, *Proc. Natl. Acad. Sci.*, 74, 2701, 1977.
14. **Whaley, K. and Ruddy, S.**, Modulation of the alternative complement pathway by β1H globulin, *J. Exp. Med.*, 144, 1147, 1976.
15. **Law, S. K., Fearon, D. T., and Levine, R. P.** Action of the C3b-inactivator on cell-bound C3b, *J. Immunol.*, 122, 759, 1979.
16. **Theofilopoulos, A. N.**, Complement receptors: properties and functions, in *Immunopathology — VII International Symposium*, Miescher, P. A., Ed., Grune and Stratton, New York, 1977, 302.
17. **Henson, P. M.**, The adherence of leukocytes and platelets induced by fixed IgG antibody or complement, *Immunology*, 16, 107, 1969.
18. **Huber, H., Polley, M. J., Linscott, W. D., Fudenberg, H. H., and Müller-Eberhard, H. J.**, Human monocytes: distinct receptor sites for the third component of complement and for immunoglobulin G, *Science*, 162, 1281, 1968.
19. **Reynolds, H. Y., Atkinson, J. P., Newball, H. H., and Frank, M. M.**, Receptors for immunoglobulin and complement on human alveolar macrophages, *J. Immunol.*, 114, 1813, 1975.
20. **Bianco, C., Patrick, R., and Nussenzweig, V.**, A population of lymphocytes bearing a membrane receptor for antigen-antibody-complement complexes. I. Separation and characterization, *J. Exp. Med.*, 132, 702, 1970.
21. **Nelson, R. A., Jr.**, The immune-adherence phenomenon. An immunologically specific reaction between microorganisms and erythrocytes leading to enhanced phagocytosis, *Science*, 118, 733, 1953.
22. **Gelfand, M. C., Frank, M. M., and Green, I.**, A receptor for the third component of complement in the human renal glomerulus, *J. Exp. Med.*, 142, 1029, 1975.
23. **Cooper, N. R.**, Isolation and analysis of the mechanism of action of an inactivator of C4b in normal human serum, *J. Exp. Med.*, 141, 890, 1975.
24. **Frank, M. M.**, The complement system in host defense and inflammation, *Rev. Infect. Dis.*, 1, 483, 1979.
25. **Pillemer, L., Schoenberg, M. D., Blum, L., and Wurz, L.**, Properdin system and immunity. II. Interaction of the properdin system with polysaccharides, *Science*, 122, 545, 1955.
26. **Nelson, R. A., Jr.**, An alternative mechanism for the properdin system, *J. Exp. Med.*, 108, 515, 1958.
27. **Muschel, L. H.**, The antibody-complement system and properdin. A review, *Vox Sang.*, 6, 385, 1961.
28. **Ellman, L., Green, I., Judge, F., and Frank, M. M.**, *In vivo* studies in C4-deficient guinea pigs, *J. Exp. Med.*, 134, 162, 1971.
29. **Frank, M. M., May, J., Gaither, T., and Ellman, L.**, *In vitro* studies of complement function in sera of C4-deficient guinea pigs, *J. Exp. Med.*, 134, 176, 1971.
30. **Sandberg, A. L., Osler, A. G., Shin, H. S., and Oliveira, B.**, The biologic activities of guinea pig antibodies. II. Modes of complement interaction with γ1 and γ2 immunoglobulins, *J. Immunol.*, 104, 329, 1970.
31. **Marcus, R. L., Shin, H. S., and Mayer, M. M.**, An alternate complement pathway: C-3 cleaving activity, not due to $\overline{C4,2a}$, on endotoxic lipopolysaccharide after treatment with guinea pig serum; relation to properdin, *Proc. Natl. Acad. Sci.*, 68, 1351, 1971.
32. **Götze, O. and Müller-Eberhard, H. J.**, The C3-activator system: an alternate pathway of complement activation, *J. Exp. Med.*, 134, 90S, 1971.
33. **Fearon, D. T. and Austen, K. F.**, Activation of the alternative complement pathway due to resistance of zymosan-bound amplification convertase to endogenous regulatory mechanisms, *Proc. Natl. Acad. Sci.*, 74, 1683, 1977.
34. **Schreiber, R. D. and Müller-Eberhard, H. J.**, Assembly of the cytolytic alternative pathway of complement from 11 isolated plasma proteins, *J. Exp. Med.*, 148, 1722, 1978.
35. **Fearon, D. T. and Austen, K. F.**, Activation of the alternative complement pathway with rabbit erythrocytes by circumvention of the regulatory action of endogenous control proteins, *J. Exp. Med.*, 146, 22, 1977.
36. **Kazatchkine, M. D., Fearon, D. T., and Austen, K. F.**, Human alternative complement pathway: membrane-associated sialic acid regulates the competition between B and β1H for cell-bound C3b, *J. Immunol.*, 122, 75, 1979.
37. **Fearon, D. T.**, Regulation by membrane sialic acid of β1H-dependent decay-association of amplification C3 convertase of the alternative complement pathway, *Proc. Natl. Acad. Sci.*, 75, 1971, 1978.

38. **Polhill, R. B., Jr., Newman, S. L., Pruitt, K. M., and Johnston, R. B., Jr.,** Kinetic assessment of alternative complement pathway activity in a hemolytic system. II. Influence of antibody on alternative pathway activation, *J. Immunol.*, 121, 371, 1978.
39. **Sandberg, A. L., Oliveira, B., and Osler, A. G.,** Two complement interaction sites in guinea pig immunoglobulins, *J. Immunol.*, 106, 282, 1971.
40. **Winkelstein, J. A. and Shin, H. S.,** The role of immunoglobulin in the interaction of pneumococci and the properdin pathway: evidence for its specificity and lack of requirement for the Fc portion of the molecule, *J. Immunol.*, 112, 1635, 1974.
41. **Boackle, R. J., Pruitt, K. M., and Mestecky, J.,** The interactions of human complement with interfacially aggregated preparations of human secretory IgA, *Immunochemistry*, 11, 543, 1974.
42. **Robertson, J., Caldwell, J. R., Castle, J. R., and Waldman, R. H.,** Evidence for the presence of components of the alternative (properdin) pathway of complement activation in respiratory secretions, *J. Immunol.*, 117, 900, 1976.
43. **Ishizaka, T., Sian, C. M., and Ishizaka, K.,** Complement fixation by aggregated IgE through alternate pathway, *J. Immunol.*, 108, 848, 1972.
44. **Day, N. K., Gewurz, H., Johannsen, R., Finstad, J., and Good, R. A.,** Complement and complement-like activity in lower vertebrates and invertebrates, *J. Exp. Med.*, 132, 941, 1970.
45. **Hugli, T. E. and Müller-Eberhard, H. J.,** Anaphylatoxins: C3a and C5a, *Adv. Immunol.*, 26, 1, 1978.
46. **Kolb, W. P. and Müller-Eberhard, H. J.,** The membrane attack mechanism of complement. Verification of a stable C5–9 complex in free solution, *J. Exp. Med.*, 138, 438, 1973.
47. **Kolb, W. P. and Müller-Eberhard, H. J.,** The membrane attack mechanism of complement. Isolation and subunit composition of the C5b–9 complex, *J. Exp. Med.*, 141, 724, 1975.
48. **Boyle, M. D. P., Langone, J. J., and Borsos, T.,** Studies on the terminal stages of immune hemolysis. III. Distinction between the insertion of C9 and the formation of a transmembrane channel, *J. Immunol.*, 120, 1721, 1978.
49. **Ecker, E. E. and Lopez-Castro, G.,** Complement and opsonic activities of fresh human sera, *J. Immunol.*, 55, 169, 1947.
50. **Mudd, S., McCutcheon, M., and Lucké, B.,** Phagocytosis, *Physiol. Rev.*, 14, 210, 1934.
51. **Nagura, H., Asai, J., Katsumata, Y., and Kojima, K.,** Role of electric surface charge of cell membrane in phagocytosis, *Acta Pathol. Jpn.*, 23, 279, 1973.
52. **van Oss, C. J. and Gillman, C. F.,** Phagocytosis as a surface phenomenon. I. Contact angles and phagocytosis of nonopsonized bacteria, *J. Reticuloendothelial Soc.*, 12, 283, 1972.
53. **Ehlenberger, A. G. and Nussenzweig, V.,** The role of membrane receptors for C3b and C3d in phagocytosis, *J. Exp. Med.*, 145, 357, 1977.
54. **Wood, W. B., Jr., McLeod, C., and Irons, E. N.,** Studies on the mechanism of recovery in pneumococcal pneumonia. III. Factors influencing the phagocytosis of pneumococci in the lung during sulfonamide therapy, *J. Exp. Med.*, 84, 377, 1946.
55. **Wood, W. B., Jr., Smith, M. R., and Watson, B.,** Studies on the mechanism of recovery in pneumococcal pneumonia. IV. The mechanism of phagocytosis in the absence of antibody, *J. Exp. Med.*, 84, 387, 1946.
56. **Lerner, E. M.,** Phagocytosis of bacteria in the absence of antibody and the effect of physical surface. A reinvestigation of "surface phagocytosis," *J. Exp. Med.*, 104, 233, 1956.
57. **Guckian, J. C., Christensen, G. D., and Fine, D. P.,** The role of opsonins in recovery from experimental pneumococcal pneumonia, *J. Infect. Dis.*, 142, 175, 1980.
58. **Wright, A. E. and Douglas, S. R.,** An experimental investigation of the role of the blood fluids in connection with phagocytosis, *Proc. R. Soc. London Ser. B*, 72, 357, 1903.
59. **Miler, I.,** Specific and non-specific opsonins, *Curr. Topics Microbiol. Immunol.*, 51, 63, 1970.
60. **Peterson, P. K., Verhoef, J., Schmeling, D., and Quie, P. G.,** Kinetics of phagocytosis and bacterial killing by human polymorphonuclear leukocytes and monocytes, *J. Infect. Dis.*, 136, 502, 1977.
61. **Wright, A. E. and Douglas, S. R.,** Further observations on the role of the blood fluids in connection with phagocytosis, *Proc. R. Soc. London Ser. B*, 73, 128, 1904.
62. **Ward, H. K. and Enders, J. F.,** An analysis of the opsonic and tropic action of normal and immune serum based on experiments with the pneumococcus, *J. Exp. Med.*, 57, 527, 1933.
63. **Boyden, S. V., North, R. J., and Faulkner, S. M.,** Complement and the activity of phagocytes, in *Ciba Found. Symp. on Complement*, Wolstenholme, G. E. W. and Knight, J., Eds., Little, Brown, Boston, 1965, 190.
64. **Griffin, F. M., Jr. Griffin, J. A., Leider, J. E., and Silverstein, S. C.,** Studies on the mechanism of phagocytosis. I. Requirements for circumferential attachment of particle-bound ligands to specific receptors on the macrophage plasma membrane, *J. Exp. Med.*, 142, 1263, 1975.
65. **Stossel, T. P.,** Phagocytosis: recognition and ingestion, *Semin. Hematol.*, 12, 83, 1975.

66. **Quie, P. G., White, J. G., Holmes, B., and Good, R. A.**, *In vitro* bactericidal capacity of human polymorphonuclear leukocytes: diminished activity in chronic granulomatous disease of childhood, *J. Clin. Invest.*, 46, 668, 1967.
67. **Root, R. K., Rosenthal, A. S., and Balestra, D. J.**, Abnormal bactericidal, metabolic, and lysosomal functions of Chediak-Higashi syndrome leukocytes, *J. Clin. Invest.*, 51, 649, 1972.
68. **Baehner, R. L. and Nathan, D. G.**, Quantitative nitroblue tetrazolium test in chronic granulomatous disease, *N. Engl. J. Med.*, 278, 971, 1968.
69. **Karnovsky, M. L.**, Metabolic basis of phagocytic activity, *Physiol. Rev.*, 42, 143, 1962.
70. **Curnutte, J. T. and Babior, B. M.**, Biological defense mechanisms. The effect of bacteria and serum on superoxide production by granulocytes, *J. Clin. Invest.*, 53, 1662, 1974.
71. **Allen, R. C.**, Evaluation of serum opsonic capacity by quantitating the initial chemiluminescent response from phagocytizing polymorphonuclear leukocytes, *Infect. Immun.*, 15, 828, 1977.
72. **Scribner, D. J. and Fahrney, D.**, Neutrophil receptors for IgG and complement: their roles in the attachment and ingestion phases of phagocytosis, *J. Immunol.*, 116, 892, 1976.
73. **Saba, T. M. and DiLuzio, N. R.**, Comparative evaluation of the influence of opsonins on hepatic, splenic and pulmonary phagocytosis, *Proc. Soc. Exp. Biol. Med.*, 125, 630, 1967.
74. **Jenkin, C. R. and Rowley, D.**, Opsonins as determinants of survival in intraperitoneal infections of mice, *Nature (London)*, 184, 474, 1959.
75. **Koenig, M. G., Melly, M. A., Goodman, J. S., and Rogers, D. E.**, Serum factors and the reticuloendothelial uptake of *Staphylococcus aureus*. I. The role of whole serum, *J. Immunol.*, 100, 516, 1968.
76. **Gigli, I. and Nelson, R. A., Jr.**, Complement dependent immune phagocytosis. I. Requirements for C'1, C'4, C'2, C'3, *Exp. Cell Res.*, 51, 45, 1968.
77. **Mantovani, B., Rabinovitch, M., and Nussenzweig, V.**, Phagocytosis of immune complexes by macrophages. Different roles of the macrophage receptor sites for complement (C3) and for immunoglobulin (IgG), *J. Exp. Med.*, 135, 780, 1972.
78. **Messner, R. P. and Jelinek, J. G.**, Inhibition of γG-mediated *in vitro* phagocytosis by the C1q component of complement, *Clin. Immunol. Immunopathol.*, 1, 203, 1973.
79. **Mantovani, B.**, Different roles of IgG and complement receptors in phagocytosis by polymorphonuclear leukocytes, *J. Immunol.*, 115, 15, 1975.
80. **Griffin, F. M., Jr., Bianco, C., and Silverstein, S. C.**, Modification of the complement receptor upon macrophage activation, *J. Cell Biol.*, 63, 123a, 1974.
81. **Griffin, F. M., Jr., Bianco, C., and Silverstein, S. C.**, Characterization of the macrophage receptor for complement and demonstration of its functional independence from the receptor for the Fc portion of immunoglobulin G, *J. Exp. Med.*, 141, 1269, 1975.
82. **Griffin, F. M., Jr. and Silverstein, S. C.**, Segmental response of the macrophage plasma membrane to a phagocytic stimulus, *J. Exp. Med.*, 139, 323, 1974.
83. **Munthe-Kaas, A. C., Kaplan, G., and Seljelid, R.**, On the mechanism of internalization of opsonized particles by rat Küpffer cells *in vitro*, *Exp. Cell Res.* 103, 201, 1976.
84. **Kaplan, G.**, Differences in the mode of phagocytosis with Fc and C3 receptors in macrophages, *Scand. J. Immunol.*, 6, 797, 1977.
85. **Sterzl, J.**, The opsonic activity of complement in sera without antibody, *Folia Microbiol.*, 8, 240, 1963.
86. **Midtvedt, T. and Trippestad, A.**, Opsonizing and bactericidal effects of sera from gnotobiotic and conventionalized rats on ^{32}P-labelled *E. coli*, *Acta Pathol. Microbiol. Scand. Ser. B*, 78, 1, 1970.
87. **Williams, R. C., Jr. and Quie, P. G.**, Opsonic activity of agammaglobulinemic human sera, *J. Immunol.*, 106, 51, 1971.
88. **Jasin, H. E.**, Human heat labile opsonins: evidence for their mediation via the alternate pathway of complement activation, *J. Immunol.*, 109, 26, 1972.
89. **Guckian, J. C., Christensen, W. D., and Fine, D. P.**, Evidence for quantitative variability of bacterial opsonic requirements, *Infect. Immun.*, 19, 822, 1978.
90. **Root, R. K., Ellman, L., and Frank, M. M.**, Bactericidal and opsonic properties of C4-deficient guinea pig serum, *J. Immunol.*, 109, 477, 1972.
91. **Clark, R. A. and Klebanoff, S. J.**, Role of the classical and alternative complement pathways in chemotaxis and opsonization: studies of human serum deficient in C4, *J. Immunol.*, 120, 1102, 1978.
92. **Austen, K. F.**, Discussion, in *Ciba Found. Symp. on Complement*, Wolstenholme, G. E. W. and Knight, J., Eds., Little, Brown, Boston, 1965, 217.
93. **Glovsky, M. M., Alenty, A., and Ghekiere, L.**, Requirement of human C2 and C3 for opsonization and bactericidal activity with human neutrophils, in *Clinical Aspects of the Complement System*, Opferkuch, W., Rother, K., and Schultz, D. R., Eds., Georg Thieme Verlag, Stuttgart, 1978, 152.

94. **Babior, B. M.**, Oxygen-dependent microbial killing by phagocytes, *N. Engl. J. Med.*, 298, 659, 721, 1978.
95. **Jenkin, C. R.**, The effect of opsonins on the intracellular survival of bacteria, *Br. J. Exp. Pathol.*, 44, 47, 1963.
96. **Glynn, A. A. and Medhurst, F. A.**, Possible extracellular and intracellular bactericidal actions of mouse complement, *Nature (London)*, 213, 608, 1967.
97. **Menzel, J., Jungfer, H., and Gemsa, D.**, Contribution of immunoglobulins M and G, complement, and properdin to the intracellular killing of *Escherichia coli* by polymorphonuclear leukocytes, *Infect. Immun.*, 19, 659, 1978.
98. **Menzel, J., Jungfer, H., and Gemsa, D.**, Amplification of the intracellular killing of *Escherichia coli* in human polymorphonuclear leukocytes by complement, in *Clinical Aspects of the Complement System*, Opferkuch, W., Rother, K., and Schultz, D. R., Georg Thieme Verlag, Stuttgart, 1978, 150.
99. **Van Snick, J. L. and Masson, P. L.**, The effect of complement on the ingestion of soluble antigen-antibody complexes and IgM aggregates by mouse peritoneal macrophages, *J. Exp. Med.*, 148, 903, 1978.
100. **Leijh, P. C. J., van den Barselaar, M. T., van Zwet, T. L., Daha, M. R., and van Furth, R.**, Requirement of extracellular complement and immunoglobulin for intracellular killing of microorganisms by human monocytes, *J. Clin. Invest*, 63, 772, 1979.
101. **Guckian, J. C., Cavallo, T., and Fine, D. P.**, A heat-labile serum factor other than complement is required for intracellular killing of pneumococci, manuscript submitted.
102. **Nelson, R. A., Jr.**, The immune-adherence phenomenon. A hypothetical role of erythrocytes in defence against bacteria and viruses, *Proc. R. Soc. Med.*, 49, 55, 1956.
103. **van Loghem, J. J., von dem Borne, A. E. G. Kr., van der Hart, M., and Peetoom, F.**, Immune adherence and blood cell destruction, *Vox Sang.*, 12, 361, 1967.
104. **May, J. E., Kane, M. A., and Frank, M. M.**, Immune adherence by the alternate complement pathway, *Proc. Soc. Exp. Biol. Med.*, 141, 287, 1972.
105. **Frank, M. M. and Atkinson, J. P.**, Complement in clinical medicine, *Dis. Mon.*, January 1975.
106. **Eden, A., Bianco, C., and Nussenzweig, V.**, Mechanism of binding of soluble immune complexes to lymphocytes, *Cell. Immunol.*, 7, 459, 1973.
107. **Pepys, M. B.**, Role of complement in the induction of immunological responses, *Transplant. Rev.*, 32, 93, 1976.
108. **Dukor, P. and Hartmann, K. U.**, Hypothesis: bound C3 as the second signal for B-cell activation, *Cell. Immunol.*, 7, 349, 1973.
109. **Waldmann, H. and Lachmann, P. J.**, The failure to show a necessary role for C3 in the *in vitro* antibody response, *Eur. J. Immunol.*, 5, 185, 1975.
110. **Alper, C. A., Colten, H. R., Gear, J. S. S., Rabson, A. R., and Rosen, F. S.**, Homozygous human C3 deficiency. The role of C3 in antibody production, C-1s-induced vasopermeability, and cobra venom-induced passive hemolysis, *J. Clin. Invest.*, 57, 222, 1976.
111. **Möller, G. and Coutinho, A.**, Role of C'3 and Fc receptors in B-lymphocyte activation, *J. Exp. Med.*, 141, 647, 1975.
112. **Lewis, G. K., Ranken, R., and Goodman, J. W.**, Complement-dependent and -independent pathways of T cell—B cell cooperation, *J. Immunol.*, 118, 1744, 1977.
113. **Brier, A. M., Chess, L., and Schlossman, S. F.**, Human antibody-dependent cellular cytotoxicity. Isolation and identification of a subpopulation of peripheral blood lymphocytes which kill antibody-coated autologous target cells, *J. Clin. Invest.*, 56, 1580, 1975.
114. **Scornik, J. C.**, Complement-dependent immunoglobulin G receptor function in lymphoid cells, *Science*, 192, 563, 1976.
115. **Perlmann, P., Perlmann, H., and Müller-Eberhard, H. J.**, Cytolytic lymphocytic cells with complement receptor in human blood. Induction of cytolysis by IgG antibody but not by target cell-bound C3, *J. Exp. Med.*, 141, 287, 1975.
116. **Lustig, H. J. and Bianco, C.**, Antibody-mediated cell cytotoxicity in a defined system: regulation by antigen, antibody, and complement, *J. Immunol.*, 116, 253, 1976.
117. **Ghebrehiwet, B., Medicus, R. G., and Müller-Eberhard, H. J.**, Potentiation of antibody-dependent cell-mediated cytotoxicity by target cell-bound C3b, *J. Immunol.*, 123, 1285, 1979.
118. **Nevins, T. E., Kim, Y., and Michael, A. F.**, Polyanion and complement receptor of the glomerular epithelium. Relationship to pH, *Lab. Invest.*, 37, 453, 1977.
119. **Carlo, J. R., Nagle, R. B., and Shin, M. L.**, The nature of the receptor for complement (C3b) in the human renal glomerulus, *Am. J. Clin. Pathol.*, 69, 486, 1978.
120. **Gelfand, M. C., Shin, M. L., Nagle, R. B., Green, I., and Frank, M. M.**, The glomerular complement receptor in immunologically mediated renal glomerular injury, *N. Engl. J. Med.*, 295, 10, 1976.

121. **Shin, M. L., Gelfand, M. C., Nagle, R. B., Carlo, J. R., Green, I., and Frank, M. M.,** Localization of receptors for activated complement on visceral epithelial cells of the human renal glomerulus, *J. Immunol.*, 118, 869, 1977.
122. **Pettersson, E. E., Bhan, A. K., Schneeberger, E. E., Collins, A. B., Colvin, R. B., and McCluskey, R. T.,** Glomerular C3 receptors in human renal disease, *Kidney Int.*, 13, 245, 1978.
123. **Foidart, J. B., Salmon, J. P., Berthoux, F. J., and Mahieu, P.,** Binding of soluble immune complexes to human glomerular complement receptors, *Kidney Int.*, 15, 303, 1979.
124. **Cavallo, T., Thorpe, L., and Fine, D. P.,** unpublished data.
125. **Ross, G. D. and Polley, M. J.,** Specificity of human lymphocyte complement receptors, *J. Exp. Med.*, 141, 1163, 1975.
126. **Bokisch, V. A. and Sobel, A. T.,** Receptor for the fourth component of complement on human B lymphocytes and cultured human lymphoblastoid cells, *J. Exp. Med.*, 140, 1336, 1974.
127. **Ross, G. D. and Rabellino, E. M.,** Identification of a neutrophil and monocyte complement receptor (CR_3) that is distinct from lymphocyte CR_1 and CR_2 and specific for a site contained within C3bi, *Fed. Proc.*, 38, 1467, 1979.
128. **Rabellino, E. M., Ross, G. D., and Polley, M. J.,** Membrane receptors of mouse leukocytes. I. Two types of complement receptors for different regions of C3, *J. Immunol.*, 120, 879, 1978.
129. **Reynolds, H. Y., Atkinson, J. P., Newball, H. H., and Frank, M. M.,** Receptors for immunoglobulin and complement on human alveolar macrophages, *J. Immunol.*, 114, 1813, 1975.
130. **Cochrane, C. G. and Dixon, F. J.,** Immune complex injury, in *Immunological Diseases*, 3rd ed., Samter, M., Ed., Little, Brown, Boston, 1978, 211.
131. **Ward, P. A. and Cochrane, C. G.,** Bound complement and immunologic injury of blood vessels, *J. Exp. Med.*, 121, 215, 1965.
132. **Cochrane, C. G.,** The role of immune complexes and complement in tissue injury, *J. Allerg.*, 42, 113, 1968.
133. **Jungi, T. W. and McGregor, D. D.,** Role of complement in the expression of delayed-type hypersensitivity in rats: studies with cobra venom factor, *Infect. Immun.*, 23, 633, 1979.
134. **Ward, P. A., Cochrane, C. G., and Müller-Eberhard, H. J.,** The role of serum complement in chemotaxis of leukocytes *in vitro*, *J. Exp. Med.*, 122, 327, 1965.
135. **Snyderman, R., Gewurz, H., and Mergenhagen, S. E.,** Interaction of the complement system with endotoxic lipopolysaccharide. Generation of a factor chemotactic for polymorphonuclear leukocytes, *J. Exp. Med.*, 128, 259, 1968.
136. **Snyderman, R., Phillips, J. K., and Mergenhagen, S. E.,** Biological activity of complement *in vivo*. Role of C5 in the accumulation of polymorphonuclear leukocytes in inflammatory exudates, *J. Exp. Med.*, 134, 1131, 1971.
137. **Fernandez, H. N., Henson, P. M., Otani, A., and Hugli, T. E.,** Chemotactic response to human C3a and C5a anaphylatoxins. I. Evaluation of C3a and C5a leukotaxis *in vitro* and under simulated *in vivo* conditions, *J. Immunol.*, 120, 109, 1978.
138. **Leddy, J. P., Frank, M. M., Gaither, T., Baum, J., and Klemperer, M. R.,** Hereditary deficiency of the sixth component of complement in man. I. Immunochemical, biologic, and family studies, *J. Clin. Invest.*, 53, 544, 1974.
139. **Goldstein, I. M.,** Endogenous regulation of complement (C5)-derived chemotactic activity: fine-tuning of inflammation, *J. Lab. Clin. Med.*, 93, 13, 1979.
140. **Gewurz, H., Page, A. R., Pickering, R. J., and Good, R. A.,** Complement activity and inflammatory neutrophil exudation in man, *Int. Arch. Allergy Appl. Immunol.*, 32, 64, 1967.
141. **Gallin, J. I., Clark, R. A., and Frank, M. M.,** Kinetic analysis of chemotactic factor generation in human serum via activation of the classical and alternate complement pathways, *Clin. Immunol. Immunopathol.*, 3, 334, 1975.
142. **Ward, P. A. and Hill, J. H.,** C5 chemotactic fragments produced by an enzyme in lysosomal granules of neutrophils, *J. Immunol.*, 104, 535, 1970.
143. **Venge, P. and Olsson, I.,** Cationic proteins of human granulocytes. VI. Effects on the complement system and mediation of chemotactic activity, *J. Immunol.*, 115, 1505, 1975.
144. **Snyderman, R., Shin, H. S., and Dannenberg, A. M., Jr.,** Macrophage proteinase and inflammation: the production of chemotactic activity from the fifth component of complement by macrophage proteinase, *J. Immunol.*, 109, 896, 1972.
145. **Hill, J. H. and Ward, P. A.,** C3 leukotactic factors produced by a tissue protease, *J. Exp. Med.*, 130, 505, 1969.
146. **Ward, P. A., Chapitis, J., Conroy, M. C., and Lepow, I. H.,** Generation by bacterial proteinases of leukotactic factors from human serum, and human C3 and C5, *J. Immunol.*, 110, 1003, 1973.
147. **Sandberg, A. L., Wahl, S. M., and Mergenhagen, S. E.,** Lymphokine production by C3b-stimulated B cells, *J. Immunol.*, 115, 139, 1975.

148. **Koopman, W. J., Sandberg, A. L., Wahl, S. M., and Mergenhagen, S. E.,** Interaction of soluble C3 fragments with guinea pig lymphocytes. Comparision of effects of C3a, C3b, C3c, and C3d on lymphokine production and lymphocyte proliferation, *J. Immunol.*, 117, 331, 1976.
149. **Smith, C. W., Hollers, J. C., Bing, D. H., and Patrick, R. A.,** Effects of human C1 inhibitor on complement-mediated human leukocyte chemotaxis, *J. Immunol.*, 114, 216, 1975.
150. **Rother, K.,** Leucocyte mobilizing factor: a new biological activity derived from the third component of complement, *Eur. J. Immunol.*, 2, 550, 1972.
151. **McCall, C. E., DeChatelet, L. R., Brown, D., and Lachmann, P.,** New biological activity following intravascular activation of the complement cascade, *Nature (London)*, 249, 841, 1974.
152. **Ghebrehiwet, B. and Müller-Eberhard, H. J.,** C3e: an acidic fragment of human C3 with leukocytosis-inducing activity, *J. Immunol.*, 123, 616, 1979.
153. **Fehr, J. and Jacob, H. S.,** *In vitro* granulocyte adherence and *in vivo* margination: two associated complement-dependent functions. Studies based on the acute neutropenia of filtration leukophoresis, *J. Exp. Med.*, 146, 641, 1977.
154. **Craddock, P. R., Fehr, J., Brigham, K. L., Kronenberg, R. S., and Jacob, H. S.,** Complement and leukocyte-mediated pulmonary dysfunction in hemodialysis, *N. Engl. J. Med.*, 296, 769, 1977.
155. **Weisdorf, D. J., Hammerschmidt, D. E., Jacob, H. S., and Craddock, P. R.,** Rapid *In vivo* Clearance of C5a: A Protective Mechanism Against Complement (C) — Mediated Tissue Injury, Proc. 8th Int. Complement Workshop, Key Biscayne, Fla., October 1979.
156. **Goldblum, S. E., Reed, W. P., Sopher, R., and Palmer, D. L.,** Pneumococcal-induced Pulmonary Leukostasis, Proc. 19th Interscience Conf. Antimicrob. Agents Chemother., Boston, September 1979, 1090.
157. **Sacks, T., Moldow, C. F., Craddock, P. R., Bowers, T. K., and Jacob, H. S.,** Oxygen radicals mediate endothelial cell damage by complement-stimulated granulocytes, *J. Clin. Invest.*, 61, 1161, 1978.
158. **Issekutz, A. C., Lee, K.-Y., and Biggar, W. D.,** Enhancement of human neutrophil bactericidal activity by chemotactic factors, *Infect. Immun.*, 24, 295, 1979.
159. **Grant, J. A., Dupree, E., Goldman, A. S., Schultz, D. R., and Jackson, A. L.,** Complement-mediated release of histamine from human leukocytes, *J. Immunol.*, 114, 1101, 1975.
160. **Brown, D. L.,** Complement and coagulation, *Br. J. Haematol.*, 30, 377, 1975.
161. **Martin, S. E., Breckenridge, R. T., Rosenfeld, S. I., and Leddy, J. P.,** Responses of human platelets to immunologic stimuli: independent roles for complement and IgG in zymosan activation, *J. Immunol.*, 120, 9, 1978.
162. **Humphrey, J. H.,** Haemolytic efficiency of rabbit IgG anti-Forssman antibody and its augmentation by anti-rabbit IgG, *Nature (London)*, 216, 1295, 1967.
163. **May, J. E., Green, I., and Frank, M. M.,** The alternate complement pathway in cell damage: antibody-mediated cytolysis of erythrocytes and nucleated cells, *J. Immunol.*, 109, 595, 1972.
164. **Reynolds, B. L and Rowley, D.,** Sensitization of complement resistant bacterial strains, *Nature (London)*, 221, 1259, 1969.
165. **Davis, S. D., Iannetta, A., and Wedgwood, R. J.,** Bactericidal reactions of serum, in *Biological Activities of Complement*, Ingram, D. G., Ed., S. Karger, Basel, 1972, 43.
166. **Platts-Mills, T. A. E. and Ishizaka, K.,** Activation of the alternate pathway of human complement by rabbit cells, *J. Immunol.*, 113, 348, 1974.
167. **Nelson, B. and Ruddy, S.,** Enhancing role of IgG in lysis of rabbit erythrocytes by the alternative pathway of human complement, *J. Immunol.*, 122, 1994, 1979.
168. **Schreiber, R. D., Morrison, D. C., Podack, E. R., and Müller-Eberhard, H. J.,** Bactericidal activity of the alternative complement pathway generated from 11 isolated plasma proteins, *J. Exp. Med.*, 149, 870, 1979.
169. **Rother, K., Rother, U., Petersen, K. F., Gemsa, D., and Mitze, F.,** Immune bactericidal activity of complement. Separation and description of intermediate steps, *J. Immunol.*, 93, 319, 1964.
170. **Heddle, R. J., Knop, J., Steele, E. J., and Rowley, D.,** The effect of lysozyme on the complement-dependent bactericidal action of different antibody classes, *Immunology*, 28, 1061, 1975.
171. **Sirotak, L., Inoue, K., Okada, M., and Amano, T.,** Immune bactericidal reactions by guinea-pig $\gamma 1$ and $\gamma 2$ antibodies, *Immunology*, 30, 435, 1976.
172. **Bladen, H. A., Evans, R. T., and Mergenhagen, S. E.,** Lesions in *Escherichia coli* membranes after actions of antibody and complement, *J. Bacteriol.*, 91, 2377, 1966.
173. **Inoue, K., Kinoshita, T., Okada, M., and Akiyama, Y.,** Release of phospholipids from complement-mediated lesions on the surface structure of *Escherichia coli*, *J. Immunol.*, 119, 65, 1977.
174. **Inoue, K., Tanigawa, Y., Takubo, M., Satani, M., and Amano, T.,** Quantitative studies on immune bacteriolysis. II. The role of lysozyme in immune bacteriolysis, *Biken J.*, 2, 1, 1959.

175. **Wardlaw, A. C.**, The complement-dependent bacteriolytic activity of normal human serum. I. The effect of pH and ionic strength and the role of lysozyme, *J. Exp. Med.*, 115, 1231, 1962.
176. **Inoue, K., Takamizawa, A., Kurimura, T., and Yonemasu, K.**, Studies on the immune bacteriolysis. XIII. Leakage of enzymes from *Escherichia coli* during immune bacteriolysis, *Biken J.*, 11, 193, 1968.
177. **Inoue, K., Yonemasu, K., Takamizawa, A., and Amano, T.**, Studies on the immune bacteriolysis. XIV. Requirement of all nine components of complement for immune bacteriolysis, *Biken J.*, 11, 203, 1968.
178. **Muschel, L. H.**, Serum bactericidal actions, *Ann. N.Y. Acad. Sci.*, 88, 1265, 1969.
179. **Donaldson, D. M., Roberts, R. R., Larsen, H. S., and Tew, J. G.**, Interrelationship between serum beta-lysin, lysozyme, and the antibody-complement system in killing *Escherichia coli*, *Infect. Immun.*, 10, 657, 1974.
180. **Davis, S. D. and Wedgwood, R. J.**, Kinetics of the bactericidal action of normal serum on Gram-negative bacteria, *J. Immunol.*, 95, 75, 1965.
181. **Reynolds, B. L. and Pruul, H.**, Sensitization of complement-resistant smooth Gram-negative bacterial strains, *Infect. Immun.*, 3, 365, 1971.
182. **Reynolds, B. L., Rother, U. A., and Rother, K. O.**, Interaction of complement components with a serum-resistant strain of *Salmonella typhimurium*, *Infect. Immun.*, 11, 944, 1975.
183. **Mayer, M. M.**, Complement and complement fixation, in *Experimental Immunochemistry*, Kabat, E. A. and Mayer, M. M., Eds., Charles C Thomas, Springfield, Ill., 1961, 133.
184. **Rapp, H. J. and Borsos, T.**, *Molecular Basis of Complement Action*, Appleton-Century-Crofts, New York, 1970.
185. **Polhill, R. B., Jr., Pruitt, K. M., and Johnston, R. B., Jr.**, Kinetic assessment of alternative complement pathway activity in a hemolytic system. I. Experimental and mathematical analyses, *J. Immunol.*, 121, 363, 1978.
186. **Ruddy, S., Carpenter, C. B., Chin, K. W., Knostman, J. N., Soter, N. A., Götze, O., Müller-Eberhard, H. J., and Austen, K. F.**, Human complement metabolism: an analysis of 144 studies, *Medicine*, 54, 165, 1975.
187. **Verbrugh, H. A., van Dijk, W. C., van Erne, M. E., Peters, R., Peterson, P. K., and Verhoef, J.**, Quantitation of the third component of human complement attached to the surface of opsonized bacteria: opsonin-deficient sera and phagocytosis-resistant strains, *Infect. Immun.*, 26, 808, 1979.
188. **Cochrane, C. G., Müller-Eberhard, H. J., and Aikin, B. S.**, Depletion of plasma complement *in vivo* by a protein of cobra venom: its effect on various immunologic reactions, *J. Immunol.*, 105, 55, 1970.
189. **Fine, D. P., Marney, S. R., Jr., Colley, D. G., Sergent, J. S., and Des Prez, R. M.**, C3 shunt activation in human serum chelated with EGTA, *J. Immunol.*, 109, 807, 1972.
190. **Fine, D. P.**, Comparison of ethyleneglycoltetraacetic acid and its magnesium salt as reagents for studying alternative complement pathway function, *Infect. Immun.*, 16, 124, 1977.

Chapter 2

HUMAN COMPLEMENT DEFICIENCIES

I. CONGENITAL COMPLEMENT DEFICIENCIES

Infections, or lack thereof, encountered by people with inherited complement deficiencies provide important clues to functions of complement components and pathways in defense against infectious diseases. This chapter will review some human complement deficiencies, with special emphasis on infectious complications. Review of the entire subject of congenital complement deficiencies has recently been published.[1] Experimental infections in congenitally deficient animals will be considered in the contexts of specific microorganisms.

A. Classical Pathway Deficiencies

Deficiencies of components of the classical pathway, especially C2, appear to be the most common of all the complement deficiencies; the incidence of C2 deficiency may be as high as 1 per 10,00 persons.[2] In general, patients lacking components of the classical pathway (C1, C4, C2) have either been in good health or had some "autoimmune" disorder.[1] The latter association with autoimmune disorders appears not to represent a sampling bias.[2] Two hypotheses have been advanced to explain this association. Synthetic genes for some components are linked rather closely to HLA genes, and thus the propensity to autoimmune phenomena could reflect the immune response status of the patient and be only coincidentally related to the complement status. Alternatively, since it is known that early complement components play a role in neutralization of certain viruses (see Chapter 5), it is also possible that complement deficiencies predispose to chronic, low-grade viral infections that are in some way etiologically related to subsequent autoimmune phenomena. Jackson and colleagues[3] demonstrated impaired antibody response to immunization with a T-dependent antigen (bacteriophage $\phi\chi$ 174) in a patient with C4 deficiency. The patient's serum also showed delayed alternative pathway activation. The investigators postulated that the combination of inability to activate the classical pathway, compromised alternative pathway function, and impaired antibody response to T-dependent antigens might predispose to prolonged viral infections or immune-complex circulation.

Patients with early component deficiencies have not generally had problems with recurrent or severe bacterial infections. A few intriguing cases in the literature, however, suggest that some C2-deficient patients may be unusually prone to infection with organisms other than viruses. Glovsky et al.[4] described two children with recurrent fevers, streptococcal pharyngitis, otitis media, urinary tract infections, and *Streptococcus pneumoniae* septicemia. More recently, Newman et al.[5] have described two unrelated children with salmonellosis, pneumococcal sepsis on multiple occasions, multiple respiratory tract infections, perirectal abscesses, and periorbital cellulitis. The patient described by Sussman et al.[6] had a syndrome of anaphylactoid purpura associated with serum particles resembling mycoplasmas. This last patient was apparently improved with administration of tetracycline.[7] Interestingly enough, the patients described by Newman et al.[5] and Sussman et al.[6] also had half-normal levels of factor B. These observations raise the possibility that propensity to infection is associated with combined deficiencies of C2 *and* factor B. It must be pointed out that another patient described by Sussman et al.[6] also had half-normal levels of factor B and deficiency of C2 and was healthy. More C2-

deficient patients will have to be studied for levels of factor B before this question can be resolved.

In vitro studies have also demonstrated that C4-deficient guinea pig serum and C2-deficient human serum are in some circumstances less efficient as opsonic sources for a variety of bacteria including *Staphylococcus aureus* and *Streptococcus pneumoniae* and that C4-deficient guinea pig serum is less efficient in the serum bactericidal system.[8-11]

Agnello[1] has suggested that any association of C2-deficiency with infection may represent a minimal propensity manifest only in early childhood when other immune systems are not mature enough to compensate. Nevertheless, the freedom from serious infection of patients deficient in early complement components and C4-deficient guinea pigs argue against any critical role for the classical pathway in normal host defenses against most infectious agents.

B. Alternative Pathway Deficiencies

Absences of proteins exclusive to the alternative pathway have not yet been described. Considerably more time must be spent in looking before one can interpret the failure to find such deficiencies.

C. C3 Deficiency

Five patients have now been described with homozygous deficiency of the third component of complement, C3. One of these patients[12] had a syndrome characterized by rash, arthralgias, and fever, but had no demonstrable infection. With the possible exception of that case, the remaining four patients have had severe and recurrent pyogenic bacterial infections. The first described[13] had 14 episodes of pneumonia, with organisms including *S. pneumoniae* and *S. pyogenes*. The patient twice had meningococcal meningitis and multiple times, otitis media, paronychia, and impetigo. The second patient[14] had persistent diarrhea as an infant, growth retardation, multiple episodes of otitis media, recurrent urinary infections, and two episodes of pneumococcal sepsis. Organisms grown from the middle ear included *Haemophilus influenzae* type B and *S. pneumoniae*. The third patient described[15] had recurrent tonsillitis and pneumococcal pneumonia and five episodes of meningitis, two of which were documented to be caused by *S. pneumoniae*. The most recently described patient[16] had recurrent pneumonia, urinary tract infection, pyogenic arthritis, otitis media, and pharyngitis. All of these patients appear to have had a null or hypofunctioning gene with no functional C3.

A second category of C3 deficiency has been associated with hereditary absence of the C3b inactivator (C3b INA). Membrane-bound C3b, deposited either through spontaneous decay of C3 or through activation by either pathway, cannot therefore be inactivated by cleavage to C3c and C3d (see Chapter 1). Interaction of the persisting active C3b with factor B results in uncontrolled positive feedback on the alternative pathway (amplification loop) with further consumption of C3.[17] The first patient described with this syndrome,[18] referred to as C3 hypercatabolism, type 1, had had multiple pyogenic infections, including otitis media, mastoiditis, pneumonia, inguinal abscess, posterior auricular abscess, skin infections, and septicemia. Organisms isolated at various times included *Staphylococcus aureus, Proteus vulgaris, Pseudomonas aeruginosa, H. influenzae,* and beta hemolytic streptococci. A second patient[19] also manifested pyogenic infections, including one episode of relapsing pneumococcal meningitis, multiple ear infections, two episodes of meningococcal meningitis, and two episodes of bacterial meningitis in which no organisms were isolated.

Finally, there is a group of patients whose serum contains circulating factors that can cleave or activate C3. The resultant C3 deficiency (hypercatabolism, type 2), usually not absolute, is secondary to the presence of these circulating factors and may represent an acquired disorder. The first such patient[20] had a circulating C3-cleaving enzyme, a 6S beta pseudoglobulin, heat *labile* and magnesium dependent. Infections included one episode each of meningococcal meningitis and pneumococcal pneumonia and multiple episodes of sinusitis, pharyngitis, and bronchitis. This patient also had partial lipodystrophy.

A closely related if not identical group of patients has been described with some combination of lipodystrophy, C3 deficiency, and membranoproliferative glomerulonephritis. Some of these patients have had increased numbers of infections. The C3 deficiency, usually only partial, is associated with a serum factor that activates C3 and appears to be identical to C3 nephritic factor.[21,22] C3 nephritic factor has recently been identified as an IgG antibody specific for neoantigens on the C3bBb complex and functions to stabilize that complex in active form.[23,24] There is no satisfactory explanation for the association of lipodystrophy and complement deficiencies. However, the two generally precede by some years the onset of nephritis. Peters et al.[25,26] have suggested that C3 nephritic factor may not cause nephritis; rather, nephritic factor may lead to chronically low levels of C3, which in turn might increase susceptibility to infection and decrease clearance of antigens derived from the infectious agents. Resultant chronic circulation of antigen-antibody complexes may be the predisposing factor to nephritis. This hypothesis remains to be established. It should be pointed out that C3 nephritic factor is heat stable and apparently not the heat-labile beta pseudoglobulin described by Alper et al.[20]

It is clear that patients with deficiencies of C3, whether due to genetic deficiency or hypercatabolism of the molecule, have a remarkable incidence of recurrent severe pyogenic infections. This propensity to infection is not attributable to any abnormality of antibody production.[16,18,27] In vitro, as might be expected, C3-deficient sera demonstrate decreased bactericidal and opsonic activity.[14] Severity of disease in such patients strongly implies a critical importance to host defenses of the third component of complement. When considered in light of relative freedom from infection of patients with early component deficiencies, this propensity to infection further strengthens the notion that the alternative pathway is of major importance in normal host defenses.

D. Terminal Component Deficiencies
1. C5 Deficiency

Rosenfeld et al.[28] described the first cases of C5 deficiency. One patient had systemic lupus erythematosus and a remarkable number of infectious complications: axillary abscesses, otitis media, otitis externa, oral and vaginal moniliasis, prolonged herpes zoster, enterococcal septicemia with meningitis, multiple subcutaneous abscesses with pseudomonas and staphylococci, and femoral thrombophlebitis. The infections were not temporally related to steroid doses. A sister with low levels of C5, detectable only by hemolytic assay, had many respiratory infections and two episodes of pneumococcal pneumonia. Snyderman and colleagues[29,30] have reported twin sisters with C5 deficiency; both have had disseminated gonococcal infection, one with multiple recurrences in a single year. Another C5-deficient sister has been healthy.

2. C6 Deficiency

The first reported patient with C6 deficiency[31] presented with gonococcal septicemia and a history of having had gonococcal arthritis 13 months previously. She had had no other infectious problems. Lim et al.[32] described a boy with recurrent group Y *Neisseria meningitidis* meningitis. Lee et al.[33,34] had a C6-deficient patient with meningococcal bacteremia and meningitis; and the patient reported by Gold and McLean[35] had chronic meningococcemia. In their review of terminal component deficiencies, Petersen et al.[36] added four more cases, three with meningococcal infection.

In another family with C6 deficiency,[37] the propositus, a 23-year-old man previously in excellent health, developed group C *N. meningitidis* meningitis and sepsis. A brother had been well until age 18 when he developed recurrent group Y *N. meningitidis* meningitis. Both patients were completely deficient in C6, as was a 21-year-old sister in excellent health. Family members heterozygous for C6 deficiency had half normal levels but no history of abnormal propensity to infection.

3. C7 Deficiency

Of thirteen individuals described with C7 deficiency,[36,38-43] five have had neisserial infections. The patient described by Lee et al.[33,41] presented with recurrent disseminated gonococcemia and past history of meningococcal meningitis as a child. A homozygous-deficient sibling had also had meningococcal meningitis as a child; her 8-year-old son, also lacking C7, was healthy.

4. C8 Deficiency

The first case of C8 deficiency described[44] had had three episodes of gonococcemia, one of which proved especially persistent. Of the subsequently described patients,[29,36,45,46] two have had meningococcal infections[36] and one has developed gonococcal endocarditis.[29]

Another patient with C8 deficiency[37] had had one episode of meningitis during her teenage years; the organism was not known. She presented at age 43 with a second episode of meningitis associated with purpura. No organisms were isolated and the patient would not participate in further studies to allow identification of the illness as meningococcal meningitis.

5. Summary

Thus, of some 47 patients with deficiencies of C3, C5, C6, C7, and C8, at least 24 have had neisserial infections (Table 1), meningococcal (18 patients), or gonococcal (5 patients), or both (1 patient). It seems less and less likely that sampling bias could explain this putative association between terminal component deficiencies and neisserial infections.[36] Furthermore, there has been little evidence that deficiencies of terminal components are linked to the HLA system; thus mere association of the deficiencies and any tendency to infection seems unlikely.

Assuming, then, a real association of terminal component deficiencies and neisserial infections, the most likely explanation is that such patients lack serum bactericidal activity for neisseria and that this deficit is important to defense against neisserial infections. Goldschneider et al.[47] established some years ago the association of bactericidal antibodies with resistance to meningococcal infection. Petersen et al.[44] clearly demonstrated a lack of bactericidal activity of C8-deficient serum against *N. gonorrhoeae*. The same bactericidal defect can be demonstrated against *N. meningitidis*.[37]

Many patients with terminal component deficiencies, however, have no problems

Table 1
NEISSERIAL INFECTIONS REPORTED IN PATIENTS WITH TERMINAL COMPLEMENT COMPONENT DEFICIENCIES

Deficient factor	Number of patients reported	Number of patients with neisserial infections				Ref.
		Total	N. meningitidis	N. gonorrhoeae	Both	
C3	5	1	1	0	0	12—16
C3bINA	2	2	2	0	0	18,19
C3 (Hypercatabolism type II)	1	1	1	0	0	20
C5	5	2	0	2	0	28—30
C6	11	9	8	1	0	31—37
C7	13	5	4	0	1	36,38—43
C8	10	4(5?)	2(3?)	2	0	29,36,37, 44—46
Total	47	24(25?)	18(19?)	5	1	

with infections; other patients go many years without problems. Presumably, the variety of host defense systems can usually compensate for deficiency of serum bactericidal activity; perhaps only when two or more of these fail do serious infections develop.

There are only tentative estimates of the incidence of complement deficiencies among patients with disseminated gonococcemia or severe meningococcal infection. However, of 27 patients with meningococcal infections seen at the University of Texas Medical Branch Hospitals in the last few years, at least two (7.4%) have had terminal complement component deficiencies.[48] Among 30 consecutive patients with systemic neisserial infection studied by Lee et al.[33,34] two (7%) had terminal component deficiency. If these estimates are anywhere near accurate, it would behoove physicians to consider hemolytic complement titrations routine in evaluation of any patient with meningococcal disease. Whether such estimates can be applied to gonococcal disease, which is seen with considerably greater frequency, remains to be established.

E. C5 Dysfunction

Miller et al.[49] described a 3-month-old infant with eczema and diarrhea; he had eight paternal relatives with eczema. Serum from the patient and other family members could not opsonize yeast (*Saccharomyces cerevesiae*, bakers' yeast) normally. Though data were not shown, the authors mentioned deficiency in opsonization of *Staphylococcus aureus* also. Clinical disease and opsonic defects were corrected by infusion of fresh plasma. Subsequently, Jacobs and Miller[50] described two brothers with syndromes of diarrhea and skin rash in infancy. The first of the siblings also had pneumonia and colitis. These patients and several family members had the same yeast opsonizing defect, which was restored in vitro by addition of semipurified C5. No tests were conducted of opsonization of organisms actually causing disease in these children. In a subsequent editorial,[51] it was suggested that these patients had Leiner's disease, which was in some critical way related to the subtle dysfunction of C5.

In another group of patients (at least one of whom had recurrent staphylococcal and Gram-negative bacterial infections), C5 dysfunction was demonstrated and appeared to be similar to that described in the children with Leiner's syndrome.[52,53]

Tests for immune adherence, immunochemical and functional C5 levels, CH50 determinations, and opsonization of pneumococci were all completely normal. These investigators also demonstrated that normal human plasma stored 10 days at 4°C demonstrated a progressive impairment of this same yeast opsonizing activity. The defect could be restored by fresh C5 or C5-sufficient serum, but not by C3 or C5-deficient serum or normal plasma that had been stored 5 days. They showed a parallel but slower decline in ability of the plasma to opsonize human erythrocytes and to produce a chemotactic stimulus. None of these defects was associated with any decrease in C5 hemolytic activity, immune adherence titer, or C5 immunochemical activity.

Since no defect of opsonization has been demonstrated with organisms actually causing disease in these patients, it is somewhat difficult to understand the pathophysiology of this syndrome. Even more disturbing, at least one other group of investigators[54] has been unable to confirm this work even using serum from one of the described C5-dysfunctional patients. Thus the clinical significance of this syndrome remains to be established. It will be important for others to identify and describe similar patients.

F. Treatment

In general, patients with deficiencies of complement components have responded appropriately to antimicrobial agents. More specific treatment would hopefully be aimed at correction of the deficiency, in a manner analogous to correction of coagulation component deficiencies. This goal has been achieved in a least one circumstance; Ziegler et al.[55] have reported treatment of a patient with C3b INA deficiency (C3 hypercatabolism, type 1) by infusion of purified C3b INA. Infusion led to correction of all complement abnormalities and clinical improvement. Others have corrected clinical defects with transfusion of fresh plasma.[12] Such transfusion, however, carries at least the theoretical risk of immunization of the recipient.[12,56]

II. ACQUIRED COMPLEMENT DEFICIENCIES

Complement deficiencies acquired in the course of complement activation (e.g., by circulating immune complexes or by endotoxin) may indeed predispose patients to subsequent infection. Investigators have spoken of "consumptive opsinopathies" as a syndrome in which intense complement consumption impairs host defenses.[57] Bjornson and colleagues[58] have demonstrated an association of hypocomplementemia and sepsis in burned patients, though hypocomplementemia could not be correlated with any functional opsonizing defect. And children with the nephrotic syndrome, who are liable to serious infections, have impaired complement function, including opsonization and generation of chemotactic factors.[59] Chemotherapy for leukemia may produce hypocomplementemia and opsonic and bactericidal defects.[60] However, because these clinical situations are extraordinarily complex, interpretation is difficult. It has been hard to know whether the infections produced hypocomplementemia or vice versa or indeed whether there was any causal relation. Some of these and similar studies will be further considered in discussions of specific syndromes and organisms.

A. Sickle Cell Disease

Sickle cell anemia, an inherited hemoglobinopathy, is associated with a predisposition to serious systemic infection, especially with encapsulated bacteria such as *Streptococcus pneumoniae* and *H. influenzae*.[61] The peak incidence of such infections is during the first 5 years of life; and infection accounts for a disproportionate number of deaths during those years. Pearson[61] attributed this problem to progressive loss of splenic function due to the inordinate load placed upon the splenic microcirculation and phagocytic capacity by the abnormal erythrocytes; he demonstrated functional asplenia by 6 months to 3 years of age in children with sickle cell disease. He suggested the asplenia resulted in impaired antibody production, impaired clearance of organisms in the blood, and decreased opsonic activity, all of which allowed otherwise transient bacteremias to develop into overwhelming sepsis.

The diminished opsonic function in sickle cell disease has been of major interest to complementologists. Winkelstein and Drachman[62] demonstrated that serum from children with sickle cell disease, though hemolytic complement was normal, failed to support phagocytosis of *S. pneumoniae* serotypes 25 or 14 by polymorphonuclear leukocytes. Their phagocytic system used 10 to 20% serum and a morphological analysis of ingestion. Johnston and colleagues[63] confirmed these observations using an assay of nitroblue tetrazolium reduction by human polymorphonuclear leukocytes ingesting type 2 pneumococci. When pneumococci were preopsonized with high concentrations of antibody, no opsonic defect could subsequently be demonstrated; but at suboptimal concentrations of antibody, serum from patients could not support maximum opsonization. Opsonic activity could be restored by addition of small (subopsonizing) amounts of normal serum but not normal serum heated to 50°C for 30 min (to inactivate factor B). EGTA-chelated serum from the patients also was a poor opsonin for zymosan, which particle bound less C3 in chelated sickle-cell serum than in chelated normal serum. All these results led Johnston and colleagues to conclude that the opsonic defect in sickle cell disease reflected dysfunction of the alternative pathway. They could not demonstrate any defect in hemolytic complement activity or in antigenic C3. Hand and King[64] found a similar defect in opsonization of *Salmonella typhimurium,* which defect may explain the peculiarly high incidence of salmonella osteomyelitis in this population.

Other investigators have tried to identify the molecular disorder in alternative pathway opsonization. With occasional exceptions,[65] results have suggested a generalized dysfunction of the alternative pathway. Koethe et al.,[66] using cobra venom factor-generated indirect lysis of guinea pig erythrocytes as an assay, demonstrated diminished function in children with sickle cell disease. Active disease ("crisis") did not affect results. Because CH50 and factor B levels were normal, the authors concluded the defect must be a partial deficiency of factor D. Also measuring the ability of serum to lyse guinea pig erythrocytes, but measuring direct rather than indirect lysis, Wilson et al.[67] confirmed the lytic dysfunction in some of the sickle cell sera tested. The majority of sera were normal. When factor B-deficient serum (produced by heating normal serum at 50°C for 15 min) was incorporated into the assay in addition to patients' sera, the defect was still demonstrated; patients' sera were able to support normal lysis in factor D-depleted serum. Thus the authors concluded that the defect lay at the level of factor B. De Ciutiis and colleagues[68] measured immunoelectrophoretic conversion of factor B and C3 by inulin. They also concluded that there was a defect in alternative pathway activation. They found a similar dysfunction in β-thalassemia, another hemoglobinopathy. At the present time, the precise reason for such dysfunction remains unexplained and any role in susceptibility to infection remains speculative.

B. Splenectomy

It has been suggested that the hyposplenism of sickle cell disease is somehow related to the serum opsonic defect. Interest in this question has been heightened by the similar increased incidence of overwhelming sepsis (especially pneumococcal) in patients whose spleens have been surgically removed.[69] Winkelstein and Lambert[70] found no impaired opsonization of type 25 pneumococci by serum from 24 children whose spleens had been removed for a variety of medical problems, ranging from thalassemia major to Hodgkin's disease to trauma. On the other hand, inulin-induced conversion of C3 was abnormal in postsplenectomy patients (most with chronic myelogenous leukemia or lymphoma) in another study.[71] Experiments in this laboratory[72] with serum from adults splenectomized for a variety of reasons would confirm the results of Winkelstein and Lambert.[70]

C. Protein-Calorie Malnutrition

Protein-calorie malnutrition, a major problem of underdeveloped countries, is associated with an array of serious infections, prominently Gram-negative bacteremia and disseminated herpes simplex virus infection. Patients may demonstrate anergy and diminished febrile responses during infections. The similarity to immunosuppressive diseases such as lymphoreticular malignancies supports experimental data implicating cell-mediated immune dysfunction as a primary problem in protein-calorie malnutrition. But Smythe et al.[73] also documented low hemolytic complement values in 61% of patients in their study. Numerous investigators have subsequently confirmed a generalized depression of complement components.[74-77]

It has generally been assumed that the hypocomplementemia reflected the general impoverishment of body protein. In experimental studies of healthy volunteers deprived of all food for 11 days, Palmblad et al.[78] measured a significant fall in C3 but not C4 levels by day 10. But protein-calorie malnutrition is complex and evaluation of patients more difficult. Suskind et al.[75] correlated low complement values with anticomplementary activity of serum and wondered whether hypocomplementemia might be purely a secondary phenomenon, perhaps induced by circulating endotoxin or immune complexes.

III. SUMMARY AND COMMENTS

Congenital complement deficiencies allow the investigator to analyze the overall effect on health of absence of a given component. Though rare, those defects allow considerable insight into the relative importance of the various proteins. Such insights are, however, dependent on multiple other variables, including epidemiology of potentially pathogenic microbes. In subsequent chapters, one can see that experimental depletion of components or limbs of the complement system has allowed dissection of the pathophysiology of a number of infections.

To the person interested in infectious diseases, review of the complement deficiencies stresses the importance of the alternative pathway and especially C3. The alternative pathway appears to occupy a primary and central spot in the host defenses against bacteria; this conclusion is strengthened by a review of experimental evaluations of interactions of bacteria and the complement system. Equally impressive, however, should be the relative freedom from infection of most patients with complement deficiencies other than C3. Clearly, such usual good health must be borne in mind in consideration of complement's role in host defenses.

From a clinical standpoint, it seems clear that the major concern of physicians should be for patients with terminal component deficiencies, who are at risk of overwhelming neisserial infection. It is not feasible to screen for component deficiencies in the public at large or even among general patient populations. But perhaps the time is here when all patients with disseminated meningococcal or gonococcal infections should be so tested.

REFERENCES

1. **Agnello, V.,** Complement deficiency states, *Medicine,* 57, 1, 1978.
2. **Lachmann, P. J.,** Complement deficiency, infection and the rheumatic diseases, in *Infection and Immunology in the Rheumatic Diseases,* Dumonde, D.C., Ed., Blackwell Scientific, Oxford, 1976, 445.
3. **Jackson, C. G., Ochs, H. D., and Wedgwood, R. J.,** Immune response of a patient with deficiency of the fourth component of complement and systemic lupus erythematosus, *N. Engl. J. Med.,* 300, 1124, 1979.
4. **Glovsky, M. M., Opelz, G., and Terasaki, P. I.,** Genetic, opsonic, and bactericidal studies in a C2 deficient family, *Clin. Res.,* 24, 327A, 1976.
5. **Newman, S. L., Vogler, L. B., Feigin, R. D., and Johnston, R. B., Jr.,** Recurrent septicemia associated with congenital deficiency of C2 and partial deficiency of factor B and the alternative complement pathway, *N. Engl. J. Med.,* 299, 290, 1978.
6. **Sussman, M., Jones, J. H., Almeida, J. D., and Lachmann, P. J.,** Deficiency of the second component of complement associated with anaphylactoid purpura and presence of mycoplasma in the serum, *Clin. Exp. Immunol.,* 14, 531, 1973.
7. **Lachmann, P. J.,** Genetic deficiencies of the complement system, *Boll. Ist Sieroter. Milan.,* 53, 195, 1974.
8. **Root, R. K., Ellman, L., and Frank, M. M.,** Bactericidal and opsonic properties of C4-deficient guinea pig serum, *J. Immunol.,* 109, 477, 1972.
9. **Johnston, R. B., Jr., Klemperer, M. R., Alper, C. A., and Rosen, F. S.,** The enhancement of bacterial phagocytosis by serum. The role of complement components and two cofactors, *J. Exp. Med.,* 129, 1275, 1969.
10. **Repine, J. E., Clawson, C. C., and Friend, P. S.,** Influence of a deficiency of the second component of complement on the bactericidal activity of neutrophils *in vitro, J. Clin. Invest.* 59, 802, 1977.
11. **Friend, P., Repine, J. E., Kim, Y., Clawson, C. C., and Michael, A. F.,** Deficiency of the second component of complement (C2) with chronic vasculitis, *Ann. Intern. Med.,* 83, 813, 1975.
12. **Osofsky, S. G., Thompson, B. H., Lint, T. F., and Gewurz, H.,** Hereditary deficiency of the third component of complement in a child with fever, skin rash and arthralgias, and response to whole blood transfusion, *J. Pediatr.,* 90, 180, 1977.
13. **Alper, C. A., Colten, H. R., Rosen, F. S., Rabson, A. R., MacNab, G. M., and Gear, J. S. S.,** Homozygous deficiency of C3 in a patient with repeated infections, *Lancet,* 2, 1179, 1972.
14. **Ballow, M., Shira, J. E., Harden, L., Yang, S. Y., and Day, N. K.,** Complete absence of the third component of complement in man, *J. Clin. Invest.,* 56, 703, 1975.
15. **Grace, H. J., Brereton-Stiles, G. G., Vos, G. H., and Schonland, M.,** A family with partial and total deficiency of complement C3, *S. Afr. Med. J.,* 50, 139, 1976.
16. **Davis, A. E., III, Davis, J. S., IV, Rabson, A. R., Osofsky, S. G., Colten, H. R., Rosen, F. S., and Alper, C. A.,** Homozygous C3 deficiency: detection of C3 by radioimmunoassay, *Clin. Immunol. Immunopathol.,* 8, 543, 1977.
17. **Abramson, N., Alper, C. A., Lachmann, P. J., Rosen, F. S., and Jandl, J. H.,** Deficiency of C3 inactivator in man, *J. Immunol.,* 107, 19, 1971.
18. **Alper, C. A., Abramson, N., Johnston, R. B., Jr., Jandl, J. H., and Rosen, F. S.,** Increased susceptibility to infection associated with abnormalities of complement-mediated functions and of the third component of complement (C3), *N. Engl. J. Med.,* 282, 349, 1970.
19. **Thompson, R. A. and Lachmann, P. J.,** A second case of human C3b inhibitor (KAF) deficiency, *Clin. Exp. Immunol.,* 27, 23, 1977.
20. **Alper, C. A., Bloch, K. J., and Rosen, F. S.,** Increased susceptibility to infection in a patient with type II essential hypercatabolism of C3, *N. Engl. J. Med.,* 288, 601, 1973.

21. **Sissons, J. G. P., West, R. J., Fallows, J., Williams, D. G., Boucher, B. J., Amos, N., and Peters, D. K.**, The complement abnormalities of lipodystrophy, *N. Engl. J. Med.*, 294, 461, 1976.
22. **Ipp, M. M., Minta, J. O., and Gelfand, E. W.**, Disorders of the complement system in lipodystrophy, *Clin. Immunol. Immunopathol.*, 7, 281, 1977.
23. **Daha, M. R., Austen, K. F., and Fearon, D. T.**, C3 nephritic factor (C3NeF): heterogeneity, polypeptide chain structure and antigenic reactivity, *J. Immunol.*, 120, 1769, 1978.
24. **Schreiber, R. D. and Müller-Eberhard, H. J.**, Nephritic factor (TA): a homogeneous immunoglobulin directed toward the complex of C3 and factor B of human complement, *J. Immunol.*, 120, 1796, 1978.
25. **Peters, D. K., Charlesworth, J. A., Sissons, J. G. P., Williams, D. G., Boulton-Jones, J. M., Evans, D. J., Kowilsky, O., and Morel-Maroger, L.**, Mesangiocapillary nephritis, partial lipodystrophy, and hypocomplementaemia, *Lancet*, 2, 535, 1973.
26. **Peters, D. K. and Williams, D. G.**, Complement and mesangiocapillary glomerulonephritis: role of complement deficiency in the pathogenesis of nephritis, *Nephron*, 13, 189, 1974.
27. **Alper, C. A., Colten, H. R., Gear, J. S. S., Rabson, A. R., and Rosen, F. S.**, Homozygous human C3 deficiency. The role of C3 in antibody production, C-1S-induced vasopermeability, and cobra venom-induced passive hemolysis, *J. Clin. Invest.*, 57, 22, 1976.
28. **Rosenfeld, S. I., Kelly, M. E., and Leddy, J. P.**, Hereditary deficiency of the fifth component of complement in man. I. Clinical, immunochemical, and family studies, *J. Clin. Invest.*, 57, 1626, 1976.
29. **Snyderman, R., Pike, M. C., McCarty, G. A., and Ward, F. E.**, Isolated deficiencies of the fifth and eighth components of complement (C) in two families: clinical, genetic and biological correlations, *J. Immunol.*, 120, 1799, 1978.
30. **Snyderman, R., Durack, D. T., McCarty, G. A., Ward, F. E., and Meadows, L.**, Deficiency of the fifth component of complement in human subjects. Clinical, genetic and immunologic studies in a large kindred, *Am. J. Med.*, 67, 638, 1979.
31. **Leddy, J. P., Frank, M. M., Gaither, T., Baum, J., and Klemperer, M. R.**, Hereditary deficiency of the sixth component of complement in man. I. Immunochemical, biologic, and family studies, *J. Clin. Invest.*, 53, 544, 1974.
32. **Lim, D., Gewurz, A., Lint, T. F., Ghaze, M., Sepheri, B., and Gewurz, H.**, Absence of the sixth component of complement in a patient with repeated episodes of meningococcal meningitis, *J. Pediatr.*, 89, 42, 1976.
33. **Lee, T. J., Schmoyer, A., Snyderman, R., Yount, W. J., and Sparling, P. F.**, Familial deficiencies of the sixth and seventh components of complement associated with bacteremic *Neisseria* infections, in *Immunobiology of Neisseria gonorrhoeae*, Brooks, G. F., Gotschlich, E. C., Holmes, K. K., Sawyer, W. D., and Young, F. E., Eds., American Society for Microbiology, Washington, D. C., 1978, 204.
34. **Lee, T. J., Snyderman, R., Patterson, J., Rauchbach, A. S., Folds, J. D., and Yount, W. J.**, *Neisseria meningitidis* bacteremia in association with deficiency of the sixth component of complement, *Infect. Immun.*, 24, 656, 1979.
35. **Gold, R. and McLean, R. H.**, Absence of sixth component of complement (C6) in a child with chronic meningococcemia, *Pediatr. Res.*, 12, 480, 1978.
36. **Petersen, B. H., Lee, T. J., Snyderman, R., and Brooks, G. F.**, *Neisseria meningitidis* and *Neisseria gonorrhoeae* bacteremia associated with C6, C7, or C8 deficiency, *Ann. Intern. Med.*, 90, 917, 1979.
37. **Fine, D. P. Davidson, J. E., Glass, D., and Griffiss, J. M.**, Evidence of multiple host factors in the predisposition to disseminated *Neisseria meningitidis* infections associated with deficiencies of C6 and C8, manuscript in preparation.
38. **Hannema, A. J., Pondman, K. W., Döhmann, U., Gadner, H., and Dooren, L. J.**, C7 deficiency in man, *Protides Biol. Fluids*, 22, 581, 1975.
39. **Boyer, J. T., Gall, E. P., Norman, M. E., Nilsson, U. R., and Zimmerman, T. S.**, Hereditary deficiency of the seventh component of complement, *J. Clin. Invest.*, 56, 905, 1975.
40. **Delage, J. M., Bergeron, P., Simard, J., Lehner-Netsch, G., and Prochazka, E.**, Hereditary C7 deficiency. Diagnosis and HLA studies in a French-Canadian family, *J. Clin. Invest.*, 60, 1061, 1977.
41. **Lee, T. J., Utsinger, P. D., Snyderman, R., Yount, W. J., and Sparling, P. F.**, Familial deficiency of the seventh component of complement associated with recurrent bacteremic infections due to *Neisseria*, *J. Infect. Dis.*, 138, 359, 1978.
42. **Nemerow, G. R., Gewurz, H., Osofsky, S. G., and Lint, T. F.**, Inherited deficiency of the seventh component of complement associated with nephritis, *J. Clin. Invest.*, 61, 1602, 1978.

43. **Zeitz, H. J., Miller, G. W., Ali, M. A., and Lint, T. F.,** Deficiency of C7 with systemic lupus erythematosus (SLE) and solubilization of immune complexes in complement (C)-deficient sera, *Clin. Res.,* 26, 716A, 1978.
44. **Petersen, B. H., Graham, J. A., and Brooks, G. F.,** Human deficiency of the eighth component of complement. The requirement of C8 for serum *Neisseria gonorrhoeae* bactericidal activity, *J. Clin. Invest.,* 57, 283, 1976.
45. **Jasin, H. E.,** Absence of the eighth component of complement in association with systemic lupus erythematosus-like disease, *J. Clin. Invest.,* 60, 709, 1977.
46. **Giraldo, G., Degos, L., Beth, E., Sasportes, M., Marcelli, A., Gharbi, R., and Day, N. K.,** C8 deficiency in a family with xeroderma pigmentosum, *Clin. Immunol. Immunopathol.,* 8, 377, 1977.
47. **Goldschneider, I., Gotschlich, E. C., and Artenstein, M. S.,** Human immunity to the meningococcus. I. The role of humoral antibodies, *J. Exp. Med.,* 129, 1307, 1969.
48. **Fine, D. P.,** unpublished observations.
49. **Miller, M. E., Seals, J., Kaye, R., and Levitsky, L. C.,** A familial, plasma-associated defect of phagocytosis. A new cause of recurrent bacterial infections, *Lancet,* 2, 60, 1968.
50. **Jacobs, J. C. and Miller, M. E.,** Fatal familial Leiner's disease. A deficiency of the opsonic activity of serum complement, *Pediatrics,* 49, 225, 1972.
51. **Miller, M. E. and Koblenzer, P. J.,** Leiner's disease and deficiency of C5, *J. Pediatr.,* 80, 879, 1972.
52. **Miller, M. E. and Nilsson, U. R.,** A familial deficiency of the phagocytosis-enhancing activity of serum related to a dysfunction of the fifth component of complement (C5), *N. Engl. J. Med.,* 282, 354, 1970.
53. **Nilsson, U. R., Miller, M. E., and Wyman, S.,** A functional abnormality of the fifth component of complement (C5) from human serum of individuals with a familial opsonic defect, *J. Immunol.,* 112, 1164, 1974.
54. **Rosenfeld, S. I., Baum, J., Steigbigel, R. T., and Leddy, J. P.,** Hereditary deficiency of the fifth component of complement in man. II. Biological properties of C5-deficient human serum, *J. Clin. Invest.,* 57, 1635, 1976.
55. **Ziegler, J. B., Alper, C. A., Rosen, F. S., Lachmann, P. J., and Sherington, L.,** Restoration by purified C3b inactivator of complement-mediated function *in vivo* in a patient with C3b inactivator deficiency, *J. Clin. Invest.,* 55, 668, 1975.
56. **Lachmann, P. J.,** C6-deficiency in rabbits, *Protides Biol. Fluids,* 17, 301, 1970.
57. **Alexander, J. W., McClellan, M. A., Ogle, C. K., and Ogle, J. D.,** Consumptive opsinopathy: possible pathogenesis in lethal and opportunistic infections, *Ann. Surg.,* 184, 672, 1976.
58. **Bjornson, A. B., Altemeier, W. A., and Bjornson, H. S.,** The septic burned patient. A model for studying the role of complement and immunoglobulins in opsonization of opportunist micro-organisms, *Ann. Surg.,* 189, 515, 1979.
59. **Anderson, D. C., York, T. L., Rose, G., and Smith, C. W.,** Assessment of serum factor B, serum opsonins, granulocyte chemotaxis, and infection in nephrotic syndrome of children, *J. Infect. Dis.,* 140, 1, 1979.
60. **Bullen, M. G., Fine, D. P., Guckian, J. C., and Reinarz, J. A.,** Opsonic and Bactericidal Defects in Chemotherapy-treated Acute Leukemia, Proc. 19th Interscience Conf. Antimicrobial Agents and Chemotherapy, Boston, September 1979, 372.
61. **Pearson, H. A.,** Sickle cell anemia and severe infections due to encapsulated bacteria, *J. Infect. Dis.,* 136, 25S, 1977.
62. **Winkelstein, J. A. and Drachman, R. H.,** Deficiency of pneumococcal serum opsonizing activity in sickle-cell disease, *N. Engl. J. Med.,* 279, 459, 1968.
63. **Johnston, R. B., Jr., Newman, S. L., and Struth, A. G.,** An abnormality of the alternate pathway of complement activation in sickle cell disease, *N. Engl. J. Med.,* 288, 803, 1973.
64. **Hand, W. L. and King, N. L.,** Serum opsonization of salmonella in sickle cell anemia, *Am. J. Med.,* 64, 388, 1978.
65. **Strauss, R. G., Forristal, J., and West, C. D.,** The alternative pathway of complement activation (APC) in sickle cell anemia (SS), *Pediatr. Res.,* 9, 326, 1975.
66. **Koethe, S. M., Casper, J. T., and Rodey, G. E.,** Alternative complement pathway activity in sera from patients with sickle cell disease, *Clin. Exp. Immunol.,* 23, 56, 1976.
67. **Wilson, W. A., Hughes, G. R. V., and Lachmann, P. J.,** Deficiency of factor B of the complement system in sickle cell anemia, *Br. Med. J.,* 1, 367, 1976.
68. **de Ciutiis, A. C., Peterson, C. M., Polley, M. J., and Metakis, L. J.,** Alternate pathway activation in sickle cell disease and β-thalassemia major, *J. Natl. Med. Assoc.,* 70, 503, 1978.

69. **Ellis, E. F. and Smith, R. T.,** The role of the spleen in immunity. With special reference to the post-splenectomy problem in infants, *Pediatrics,* 37, 111, 1966.
70. **Winkelstein, J. A. and Lambert, G. H.,** Pneumococcal serum opsonizing activity in splenectomized children, *J. Pediatr.,* 87, 430, 1975.
71. **de Ciutiis, A., Polley, M. J., Metakis, L. J., and Peterson, C. M.,** Immunologic defect of the alternate pathway-of-complement activation postsplenectomy: a possible relation between splenectomy and infection, *J. Natl. Med. Assoc.,* 70, 667, 1978.
72. **Pagel, M. J., Guckian, J. C., and Fine, D. P.,** unpublished observations.
73. **Smythe, P. M., Schonland, M., Brereton-Stiles, G. G., Coovadia, H. M., Grace, H. J., Loening, W. E. K., Mafoyane, A., Parent, M. A., and Vos. G. H.,** Thymolymphatic deficiency and depression of cell-mediated immunity in protein-calorie malnutrition, *Lancet,* 2, 939, 1971.
74. **Sirisinha, S., Suskind, R., Edelman, R., Charupatana, C., and Olson, R. E.,** Complement and C3-proactivator levels in children with protein-calorie malnutrition and effect of dietary treatment, *Lancet,* 1, 1016, 1973.
75. **Suskind, R., Edelman, R., Kulapongs, P., Pariyanonda, A., and Sirisinha, S.,** Complement activity in children with protein-calorie malnutrition, *Am. J. Clin. Nutrition,* 29, 1089, 1976.
76. **Hafez, M., Aref, G. H., Mehareb, S. W., Kassem, A. S., El-Tahhan, H., Rizk, Z., Mahfouz, R., and Saad, K.,** Antibody production and complement system in protein energy malnutrition, *J. Trop. Med. Hyg.,* 80, 36, 1977.
77. **Kielmann, A. A. and Curcio, L. M.,** Complement (C3), nutrition and infection, *Bull. W.H.O.,* 57, 113, 1979.
78. **Palmblad, J., Cantell, K., Holm, G., Norberg, R., Strander, H., and Sunblad, L.,** Acute energy deprivation in man: effect on serum immunoglobulins antibody response, complement factors 3 and 4, acute phase reactants and interferon-producing capacity of blood lymphocytes, *Clin. Exp. Immunol.,* 30, 50, 1977.

Chapter 3

BACTERIA

I. CLASSIFICATION

Bacteria belong to the kingdom *Procaryotae*. The reader not especially versed in microbiology may find it useful to refer to Table 1, which is adapted from *Bergey's Manual of Determinative Bacteriology*,[1] and lists bacteria pathogenic for man or otherwise referred to in this text. For convenience, the discussion of interactions of bacteria with the complement system will follow the outline of this table.

II. BACTERIAL SURFACES

Because the interaction with complement occurs at the microbial surface, it is important to review some of the information regarding that surface. This subject is, of course, worthy of extensive reviews of its own; numerous articles and books have been published.[2,3]

A. Cell Membrane

Bacteria are, like all cells, limited by a metabolically active and complex cell membrane. This lipid bilayer maintains the osmotic and ionic integrity of the cell by virtue of its molecular transport function. The membrane is also a major site of synthesis of cellular products.[4]

B. Cell Wall

Outside the relatively fragile cell membrane is the cell wall, a tough, semirigid envelope which protects the bacterium in a sometimes hostile world. The *sine qua non* of this exoskeleton is a polymer known variously as peptidoglycan, glycopeptide, mucopeptide, glycosaminopeptide, and murein. The two glycans, *N*-acetylglucosamine and *N*-acetylmuramic acid, are linked through peptide side chains so as to compose "a single, enormous macromolecule that forms a more-or-less continuous network around the cellular permeability barrier and provides the cell with a supporting structure of high tensile strength."[5] The glycan portion varies only slightly among bacteria, whereas variations in the peptide moieties contribute greatly to characteristics of bacterial strains and species.[5,6]

Many bacteria can be segregated on the basis of the Gram stain. Gram-positive organisms retain the dye crystal violet; in contrast, the dye is washed out of Gram-negative bacteria by acetone-alcohol. This seemingly trivial difference reflects profound differences in biological behavior between Gram-positive and Gram-negative bacteria. To a great extent, these differences are due to the structure of the respective cell walls.

1. Gram-Positive Bacteria

Cell walls of Gram-positive bacteria are thick, poorly defined structures, comprised mostly of peptidoglycan. The nonpeptidoglycan portion includes a wide variety of polysaccharides and teichoic acids, which are covalently bound to the peptidoglycan.[5]

Teichoic acid is the collective name for water-soluble polymers of ribitol or glycerol phosphate linked by phosphodiester bonds. Variability among teichoic acids is conferred by side chains. Glycerol teichoic acids are found intracellularly associated

Table 1
AN ORGANIZED LIST OF BACTERIAL GENERA PERTINENT TO THIS MONOGRAPH[1]

Spirochetes
  Spirochaeta
  Treponema
  Borrelia
  Leptospira
Spiral and curved bacteria
  Spirillum
  Campylobacter
Gram-negative, aerobic rods
  Pseudomonas
  Alcaligenes
  Brucella
  Bordetella
  Francisella
Gram-negative, facultatively anaerobic rods
  Enterobacteriaceae
    Escherichia
    Edwardsiella
    Citrobacter
    Salmonella
    Shigella
    Klebsiella
    Enterobacter
    Hafnia
    Serratia
    Proteus
    Yersinia
    Erwinia
  Vibrionaceae
    Vibrio
    Aeromonas
  Genera of uncertain affiliation
    Chromobacterium
    Flavobacterium
    Haemophilus
    Pasteurella
    Actinobacillus
    Streptobacillus
    Calymmatobacterium
Gram-negative, anaerobic bacteria
  Bacteroides
  Fusobacterium
Gram-negative cocci and coccobacilli (aerobes)
  Neisseria
  Moraxella
  Acinetobacter
Gram-negative cocci (anaerobes)
  Veillonella
Gram-positive cocci
  Micrococcus
  Staphylococcus
  Streptococcus
  Peptococcus
  Peptostreptococcus
Endospore-forming rods
  Bacillus
  Clostridium
Gram-positive, asporogenous rod-shaped bacteria
  Lactobacillus
  Listeria
  Erysipelothrix
Actinomycetes and related organisms
  Corynebacterium
  Propionibacterium
  Actinomyces
  Mycobacterium
  Nocardia
  Streptomyces
  Micromonosporaceae
    Micromonospora
    Thermoactinomyces
    Micropolyspora
The rickettsias
  Rickettsia
  Rochalimaea
  Coxiella
  Bartonella
  Hemobartonella
  Chlamydia
The mycoplasmas
  Mycoplasma
  Acholeplasma

with plasma membranes of all Gram-positive bacteria as well as in the cell walls. Ribitol teichoic acids are limited to cell walls. Though apparently not critical to structural integrity of the cell wall, the teichoic acids may be important in ion transport. Because the teichoic acids are highly antigenic, they contribute greatly to serological differences among Gram-positive bacteria.[7]

2. Gram-Negative Bacteria

Whereas the cell walls of Gram-positive bacteria are somewhat amorphous and composed largely of peptidoglycan, the cell walls of Gram-negative bacteria are highly structured and peptidoglycan is a relatively minor constituent. A thin peptidoglycan layer lies just outside the plasma membrane; the two may be linked in some way. Outside and covalently linked to the peptidoglycan are two layers of lipopolysaccharides and lipoproteins. These three layers and the intervening spaces can be readily demonstrated by electron microscopy.[5]

The lipopolysaccharides are major somatic antigens of Gram-negative bacteria in addition to their important biological activities. The hydrophobic lipid portion of lipopolysaccharide is called lipid A and is apparently required for viability of the organism, since no mutants lacking lipid A have ever been identified. Lipid A is composed of a D-glucosamine skeleton to which are attached 3-hydroxytetradecanoic acid and medium-chain saturated fatty acids. The polysaccharide portion of the macromolecule has two components, the relatively invariable core oligosaccharide and the highly variable O side chains. The core oligosaccharide is mostly 3-deoxy-D-manno-octulosonic acid (also called 2-keto-3-deoxyoctonic acid or KDO) with some other saccharides, especially heptose.[8]

Colonies of bacterial mutants having only core polysaccharide and lipid A in their cell walls may have a visibly rough surface that can be distinguished on the agar plate from the smooth (S) colonies of parent bacteria. Such rough (R) mutants are readily killed by serum alone or in combination with phagocytes and are generally nonpathogenic. From these observations one can infer the importance of the O side chains of lipopolysaccharides. These structures are outermost on the molecule and thus in greatest contact with the environment. They are quite diverse (though constant for a given organism) and confer antigenic and biological differences by their diversity. Pathogenicity requires their presence, and host defenses employed against Gram-negative bacteria have been suggested to be the evolutionary pressure behind the diversity.[8] Mutants may be seen with various amounts of O side chain, that is, with differing degrees of roughness. Pathogenicity can also be related to quantitative differences among O side chains of Gram-negative bacteria. Thus, the greater pathogenicity of salmonellae (compared to *Escherichia coli*) has been attributed to generally consistent differences in their O side chains, virulent salmonellae tending to have repeating polysaccharide units that are otherwise rare in nature and to which infected hosts would thus be unlikely to be "naturally" immune.[9]

C. Capsule

Exterior to the cell walls of many bacteria is a structure called the capsule or slime layer. (From the German *Kapsel* comes the designation of capsular antigens in some species as "K" antigens.)[10] This highly variable structure is porous, poorly defined, amorphous, and not visualized microscopically using standard stains. The capsular layer may be very narrow or greater in width than the rest of the cell. The major constituents of bacterial capsules are polysaccharides; proteins and lipids may also be present in relatively small amounts. The site of synthesis of these capsular materials is thought to be within the cell or perhaps at the cell membrane.[11]

Although not critical to life of the cell itself, the capsule functions as a barrier to protect the cell membrane from attack by host defense mechanisms, especially antibody, complement, and phagocytes.[11] In this sense, the capsule serves a function very much like one of the functions of the O side chains of lipopolysaccharides, as discussed above. The presence of a capsule is also associated with a smooth colonial appearance and with virulence; likewise, organisms lacking capsules (rough strains) tend to be nonvirulent. In the laboratory, smooth strains may be induced to become rough and avirulent by sequential passage on suboptimum media; repassage through optimum medium or a laboratory animal may induce reversion to the smooth, virulent form.

"Smoothness" and "roughness" are only gross extremes of a continuum. Strains of the same species may be found that produce varying amounts of capsule. Studies of these variants suggest even more strikingly the value to the microbe of a capsule. MacLeod and Krauss[12] demonstrated in a mouse model of pneumococcal infection a direct correlation between the amount of capsular polysaccharide produced and the virulence of various strains of the same serotype. Such a relationship is, of course, in addition to differences in pathogenicity conferred by qualitative differences among capsules of various organisms.[13]

D. Interactions of Host Defenses with Bacterial Surfaces

This subject will be considered in greater detail in discussions of specific bacteria. However, it seems reasonable to make a few generalizations at this point. Antibodies to most cell wall structures, by virtue of the latter's ubiquity, are common in non-immune animals and humans. Furthermore, lipopolysaccharides, peptidoglycan, and teichoic acids have all been shown in various studies to react with the alternative complement pathway, often in the absence of antibody. Thus, opsonization or bacterial lysis by the humoral immune systems would theoretically be routine — and probably is so for avirulent bacteria. The outer layers however, O side chains and capsules, serve to protect the bacterium, either by obscuring otherwise reactive cell wall components or by blocking the reaction of deposited antibody or complement with receptors on phagocytes. Thus, the value of specific immunity (e.g., immunization with pneumococcal capsular polysaccharides) consists of reestablishing an immune reaction at the cell surface.

III. SPIROCHETES

The name implies the spiral or coiled appearance of these organisms. They, like the Gram-negative bacteria, have inner peptidoglycan and outer lipopolysaccharide cell walls.[14] They have not been cultured in vitro, and thus their interactions with complement have not been so well-studied as those of other bacteria.

A. *Treponema pallidum*

This organism is the etiologic agent of syphilis. One of the serologic aids to diagnosis is the *Treponema pallidum* immobilization (TPI) test. The ability of a patient's serum to inhibit the spirochete's normal frenetic movement is evidence for antibody and thus for preceding infection. Hederstedt[15] has suggested that this immobilizing activity may require complement as well as IgM antibody.

The clinical stages of syphilis are four. Primary syphilis is characterized by a small ulcer at the site of skin or mucosal invasion by the organism. The ulcer heals and several weeks later a spirochetemia ensues, characterized by fever, diffuse skin rashes associated with cutaneous infective vasculitis, and lymphadenopathy (secondary syphilis). This phase also is self-limited and is followed by a period of disease latency that may last a few months, years, or a lifetime. If latency is terminated, the final stage, tertiary syphilis, is characterized by progressive destructive disease usually localized to a particular site or organ system (e.g., central nervous system syphilis in various forms, cardiovascular syphilis, gummata).

During the spirochetemia of secondary syphilis, some patients develop a nephrotic syndrome. Renal biopsies have shown glomerulonephritis, characterized by glomerular endothelial proliferation and swelling, electron-dense deposits subendothelially and subepithelially along the basement membrane, and fusion of the epithelial foot processes.[16] Patients with secondary syphilis almost all have demonstrable serum antibody to the spirochetes; it thus is not surprising that considerable evidence exists for an immune complex mechanism in the glomerulopathy.[17] Evidence for involvement of complement in this disease has been conflicting. Several studies have been unable to demonstrate glomerular deposits of complement components.[17,18] Several other studies, however, have demonstrated granular basement membrane deposits of C3.[19-22] Serum levels of complement activity or proteins have generally been normal,[17,19,21] though Yuceoglu and colleagues[20] noted diminished whole complement hemolytic activity in two of three infants with the nephritis of congenital syphilis (which, for these purposes, may be considered a variant of secondary syphilis). In that study, complement levels returned to normal with successful antibiotic therapy.

Initiation of antibiotic therapy for syphilis, especially secondary syphilis, may be followed in 4 to 8 hr by the Jarisch-Herxheimer reaction. This dramatic event is characterized by fever, enhanced severity of the manifestations of the syphilis itself (e.g., flare of the rash), tachycardia, and hypotension. Patients may also develop anorexia, myalgias, malaise, and chills. The reaction usually clears by 24 hr.[23] Fulford et al.[23] demonstrated a preceding fall in antitreponemal antibodies and complement (CH50, C4, C3, C6, C7, $C\bar{1}$ Inh, but not factor B) in six patients. Circulating immune complexes were not detectable. The two patients with clinically most severe reactions had the greatest declines in CH50 (160ℓ to 180 U/mℓ prior to therapy, 100 U/mℓ 2 to 3 hr after therapy, and 140 to 180 U/mℓ 6 hr after therapy); these patients also had demonstrable in vivo conversion of C3 to C3b. In contrast, Gelfand and colleagues[24] noted only minimal, if any, fall in complement in two patients with the Jarisch-Herxheimer reaction. They did however find endotoxemia during the height of the reaction and suggested a role for this component of the spirochetal cell wall.

IV. GRAM-NEGATIVE, AEROBIC RODS

A. *Pseudomonas aeruginosa*

Forsgren and Quie[25] demonstrated in vitro opsonization of strains of *Pseudomonas aeruginosa* by normal serum. This opsonization was dependent on heat-labile factors and was only partially inhibited by EGTA; thus, some opsonization could be mediated by the alternative pathway. Other studies[26] suggested a major role for the alternative pathway in opsonization of *P. aeruginosa* by normal human sera, since such opsonization required not only antibody but properdin and factor B. On the other hand, in the presence of immune IgG, opsonization could proceed independently of factor B and properdin.

In the mouse, complement was required for the clearance of pseudomonas from the lung.[27] And pseudomonas isolated from sputum of patients with cystic fibrosis or patients with chronic lung disease were coated with immunoglobulins and complement as determined by immunofluorescence.[28]

In an experimental model of *P. aeruginosa* endocarditis, serum-sensitive organisms were more rapidly cleared from the blood than were serum-resistant organisms and caused no deaths, whereas serum-resistant organisms produced a disease with 78% mortality, a high rate of active endocarditis, and large numbers of organisms in valve vegetations.[29] Nevertheless, even in this model, serum-sensitive organisms circulated in the blood for long periods, up to 2 days in 77% of the animals and up to 4 days in 33%.

In humans with *P. aeruginosa* sepsis, 91% of organisms isolated from blood cultures in one study were serum resistant; however, 79% of organisms isolated from noninfected sites were also serum resistant.[30] Because these differences were not great and because even the serum-resistant organisms were readily opsonized by antibody and complement, the authors suggested that the serum bactericidal system had a limited role in host defenses against *P. aeruginosa*.

B. *Francisella tularensis*

This organism, formerly called *Pasteurella tularensis*, is the etiologic agent of tularemia. Carlisle and Saslaw[31] measured properdin levels in volunteers exposed to aerosols of the organism. They found no changes in properdin levels with development of symptoms in vaccinated or nonvaccinated subjects. In preliminary studies, they suggested that the organisms were not sensitive in vitro to properdin. However, the sensitivity or reliability of this assay is not clear in light of current understanding of the alternative pathway and the role of properdin.

V. GRAM-NEGATIVE, FACULTATIVELY ANAEROBIC RODS

A. Enterobacteriaceae

The major Gram-negative pathogens of man fall into this family. Thus, it is not surprising that most of the research regarding Gram-negative bacteria, microbiology, and complementology, has involved these microbes, especially *Escherichia coli* and the salmonellae.

1. Escherichia coli

Roantree et al.[32,33] performed classical studies of the effect of the serum bactericidal system upon Gram-negative bacteria, especially *E. coli*. They expressed serum bactericidal activity in terms of the number of organisms killed per milliliter of undiluted serum in 2 hr and defined organisms as serum sensitive if more than 10^6 bacteria could be so killed. Bacteria were moderately sensitive if 10^4 to 10^6 could be killed, serum resistant if less than 10^4 were killed. They found that, against a given organism, bactericidal activity was remarkably uniform among some 43 patients and 36 healthy adults studied. However, there was marked variability among bacteria as to sensitivity. Among 21 bacteria isolated from blood cultures, only 4 were moderately sensitive. Among 55 isolates from stool, 34 were sensitive or moderately sensitive. These differences were obtained for specific genera as well. For example, among 13 blood isolates of *E. coli*, 1 was serum sensitive, 2 were moderately sensitive, and 10 were serum resistant. Five stool isolates were serum sensitive, ten were moderately sensitive, and four were resistant. Among bacteria causing urinary tract infections, four were sensitive, eight were moderately sensitive, and four were resistant. If relatively low concentrations of bacteria (approximately $10^3/m\ell$) were incubated in vitro with human or rabbit serum, serum-sensitive bacteria were essentially completely killed within 10 min of incubation, whereas serum-resistant organisms remained viable for more than 60 min. In vivo studies were then done; 10^4 to 10^6 bacteria were injected into rabbits and serial quantitative blood cultures obtained. Serum-sensitive organisms were essentially gone from the blood stream within 15 min, though at 24 hr there was a reappearance of bacteria in the blood. In contrast, serum-resistant organisms were never completely eliminated from the blood, though numbers were reduced by several logs initially. In occasional experiments there was complete but usually transient clearing of the blood even of serum-resistant organisms. Nevertheless, the generally slower rate of clearance of serum-resistant organisms was consistent.

The investigators were careful to point out that, because even the serum-resistant organisms were rather efficiently cleared (3 to 4 log reduction in count), "factors other than the bactericidal activity of serum play the major role in clearing the blood stream of bacteria."[33] However, the consistently greater numbers of serum-resistant bacteria at most times during experimental sepsis suggested some importance for the serum bactericidal system, especially at low concentrations of bacteria (the usual case even in severe sepsis).

Rowley[34] correlated lethality for mice with in vitro serum sensitivity of various strains of E. coli. Others have noted that organisms causing pyelonephritis are less serum sensitive than organisms causing asymptomatic bacteriuria.[35] Miller et al.[36] studied serum-sensitive and serum-resistant E. coli in an experimental model of pyelonephritis. Infection was produced by direct intrarenal inoculation of the organisms into rats. Serum-resistant bacteria were not eliminated from the kidney during 48 hr of observation, whereas serum-sensitive E. coli were almost completely killed. Cobra venom factor pretreatment of the rats abolished this elimination of serum-sensitive organisms. Vosti and Randall[37] noted that, of 293 isolates of E. coli, approximately one third were serum sensitive. In contrast, only 13% of strains causing bacteremia were highly serum sensitive. Durack and Beeson[38] studied the role of serum bactericidal activity in E. coli endocarditis. They attempted to induce endocarditis with various strains of E. coli in rabbits that were either normal or deficient in C6. Eleven percent of normal animals injected with serum-sensitive E. coli developed endocarditis. In contrast, 91% of rabbits given serum-resistant E. coli developed endocarditis 1 to 7 days later. The mean bacterial counts in vegetations from animals infected with serum-sensitive E. coli were approximately 100-fold lower (per gram of vegetation) than from animals with serum-resistant E. coli. When the serum-sensitive E. coli were injected into a second group of normal rabbits, 1 of 11 developed endocarditis; however, injection of these same organisms into C6-deficient rabbits led to endocarditis in all 5 animals studied.

As noted previously, the capsular (K) antigen of E. coli is a major virulence factor, capable of obscuring the O antigens from host immune defenses. The serology, chemistry, and genetics of E. coli K and O antigens have been recently reviewed by Orskov and colleagues.[10] These authors divided K antigens into two groups: polysaccharide (acidic) and protein (fimbrial) (Table 2). The fimbriae (pili) of K88 and K99 antigens, which should be distinguished from the common type I pili of all enterobacteriaceae, are associated with animal pathogens. Most research with the complement system has involved E. coli polysaccharide antigens. Organisms isolated from human infections (urinary tract infections, septicemia, etc.) have predominately the K antigens 1, 2, 3, 5, 12, or 13. Isolates from neonates with meningitis are almost uniformly K1.[10]

The presence of K antigen has been related to in vivo resistance to phagocytosis in mice.[39] Björksten and Kaijser[40] noted that organisms isolated from patients with asymptomatic bacteriuria generally had scant K antigen and that this attribute was associated with an increased susceptibility to opsonization by the alternative complement pathway or in dilute serum. This observation could not be related to any particular O or K antigens. In another study,[41] E. coli isolates with K antigen were poorly opsonized at serum concentrations ranging from 1% to 20%. Opsonization of K antigen-containing strains was completely inhibited at all concentrations by MgEGTA; opsonization of K antigen-deficient strains, only partially. K antigen-containing strains also consumed complement less efficiently; this consumption was inhibited partially or completely by MgEGTA. Björksten, Bortolussi, and colleagues[42,43] studied the role of K1 antigen in opsonization, serum sensitivity, and virulence. Though K1 content did not appear to affect alternative pathway activation,[42] K1-containing E. coli were poorly opsonized in MgEGTA-chelated or C2-deficient sera. In newborn rats, those K1 strains resistant to alternative pathway

Table 2
K ANTIGENS OF *E. COLI*[10]

Polysaccharide (acidic) K antigens
  K1—57
  K62
  K74
  K82—84
  K87
  K92—98
  K100
Protein (fimbrial) K antigens
  K88
  K99

opsonization were more virulent than sensitive strains. Stevens et al.[44] also found that K1-containing *E. coli* were not opsonized by the alternative pathway. Forsgren and Quie,[45] however, found that *E. coli* K12 was selectively opsonized by the alternative pathway. This finding was confirmed by Glovsky et al.[46]

It has been suggested that sensitivity to the serum bactericidal system is not related to differences in K antigen content.[39,43,47] Akiyama and Inoue[47] induced serum resistance in an originally serum-sensitive *E. coli* K12 by serial passage in medium containing antiserum and guinea pig complement. They observed associated changes in fatty acids of the plasma membrane, but it was not clear whether these changes caused development of serum resistance.

Guckian et al.[48] studied two strains of *E. coli* (not characterized as to capsule), both of which required antibody and complement for opsonization, but which varied in their ability to activate the alternative pathway. In studies of *E. coli* isolated primarily from burn wounds, Leist-Welsh and Bjornson[49] found none to be opsonized in normal serum by the alternative pathway. All organisms required an intact classical pathway, although two strains appeared not to require immunoglobulin.

Gilbert et al.[50,51] studied *E. coli* bacteremia in the squirrel monkey. Their studies demonstrated complement-dependent clearance and complement dependency of the early leukopenia.

2. Salmonellae

Salmonellae may be divided into a number of species on the basis of serology. *Salmonella typhi*, the causative organism of typhoid fever, is perhaps the most pathogenic of the species, but many others may produce similar disease. Various salmonellae may produce syndromes ranging from gastroenteritis to osteomyelitis.

Rowley[52] demonstrated that strains of salmonellae that were smooth had not lost antigens with which the serum bactericidal system reacted; rather, these antigens had become inaccessible, presumably because of obscuring lipopolysaccharide. In studies with *S. minnesota* and *S. enteritidis*, Reynolds and Rowley[53] demonstrated that dissolution of the outer cell walls by agents such as EDTA led to exposure of sites to which complement could be better fixed and thus to conversion from serum resistance to serum sensitivity. In similar studies with *S. typhimurium*, Reynolds and Pruul[54] demonstrated that treatment of a smooth strain with an agent such as tris or EDTA released lipopolysaccharides and converted the organisms to serum sensitive. Apparently, the effect of the tris or EDTA was to expose sites to which complement could fix. This effect was studied further by Reynolds et al.,[55] who demonstrated that antibody to this same strain of *S. typhimurium* induced normal fixation of guinea pig components C1, C4, C2, and C3, but poor fixation of the later components. Only when these later components were supplied in an EDTA buffer was there lysis of the bacteria.

Bjornson and Bjornson[56] suggested that the polysaccharide portion of the lipopolysaccharide activated the alternative pathway. This conclusion was based on observations that smooth salmonellae activated the alternative pathway more efficiently than did a variant that lacked O side chains and the acetylglucosamine. The variant containing only lipid A could not activate the alternative pathway. They did not correlate complement activation with lysis or opsonization.

In spite of the interactions of salmonellae with the bactericidal system, Rother and Rother[57] demonstrated normal clearance of *S. typhi* injected into C6-deficient rabbits.

In clinical studies of experimental typhoid fever, Schubart et al.[58] noted a decline in hemolytic complement activity in two of six volunteers prior to the onset of fever and bacteremia. Properdin levels declined slightly in four of the five who developed fever.

3. Shigellae

Shigellae, like salmonellae, may be serologically distinguished but produce similar syndromes of dysentery. Reed and Albright[59] studied the serum sensitivity of strains of shigella. Though bactericidal activity against a given strain was relatively consistent among various sera, the sensitivities of eight strains of shigella varied widely. Small amounts of antibody, primarily IgM, were required. Bactericidal activity was diminished in C2-deficient serum or in serum artificially depleted of factor B. C2-deficient serum depleted of factor B was totally lacking in bactericidal activity. In further studies, Reed[60] noted that strains of shigellae varied also in opsonic requirements. Serum-resistant strains (that is, shigellae not killed by the serum bactericidal system) required complement for opsonization. There were variable requirements for heat-stable opsonins. Essentially all strains were susceptible to the combined effect of 19S antibodies and heat-labile opsonins. Some strains were also opsonized by heat-stable 7S serum factors alone. At least one strain could be opsonized via the alternative complement pathway.

4. Klebsiella pneumoniae

Klebsiella pneumoniae is a highly pathogenic organism, tending especially to produce disease in individuals with embarrassed antimicrobial defenses. Classically, *K. pneumoniae* is thought of as the agent of an overwhelmingly virulent pneumonia in alcoholics. It is also a major pathogen in hospital-acquired infections, in which situations the pressure of excessive antibiotic administration is thought to be a driving force.

In early studies of *K. pneumoniae* bacteremia in rabbits, Wood and colleagues[61] demonstrated polymorphonuclear leukocytes to adhere to vascular endothelial cells and ingest organisms in the absence of demonstrable antibody. Complement studies were not done; thus any attribution of their observations to complement (especially alternative pathway) mediation is purely speculative.

Gross et al.[27] demonstrated that pulmonary clearance of *K. pneumoniae* in mice following aerosolization was not dependent on complement. Their conclusions were based on normal clearance of organisms in cobra venom factor-treated mice. However, opsonization of *K. pneumoniae* in vitro required complement in addition to immunoglobulin.[49]

5. Serratia marcescens

Once considered nonpathogenic, *Serratia marcescens* is now recognized as potentially pathogenic, but primarily in patients with impaired host defenses and after more usual pathogens have been eliminated by enthusiastic antibiotic administration.

Simberkoff and colleagues[62] carried out remarkable studies of serratia isolated from two hospitals in New York City, one of which had an outbreak of serratia infections. In the hospital with an excessive infection rate, organisms isolated from various hospital locations and body sites demonstrated a low incidence of serum sensitivity (22%) when compared with isolates from the hospital free of problems with serratia infections (58% of isolates serum sensitive). This difference correlated with a tenfold increase in the number of documented bacteremic infections with *S. marcescens* during 1 year. None of the blood culture isolates from the affected hospital was serum sensitive. The investigators, in further in vitro studies of these isolates, found bactericidal activity to be serotype specific and absorbable, as well as heat-labile. There did not appear to be strain differences in opsonization, which proceeded primarily (or at least adequately) through the alternative pathway, as evidenced by lack of inhibition of opsonization by antiserum to C1 or by EGTA.

Guckian et al.[48] studied one strain of *S. marcescens* that had at least a partial requirement for heat-labile opsonins, apparently C3. This organism activated the alternative complement pathway, even in absorbed, presumably antibody-deficient serum. The two serratia strains studied by Leist-Welsh and Bjornson[49] required complement for opsonization, but only one utilized the alternative pathway.

Traub and Kleber[63] distinguished two groups of *S. marcescens:* those killed within minutes by normal serum ("promptly serum-sensitive") and those killed only after several hours exposure to serum ("delayed serum-sensitive"). Killing of the former was delayed by chelation of serum with MgEGTA, whereas chelation did not further delay serum sensitivity of the latter. The investigators concluded that the classical pathway was the primary mediator of serum killing of serratia; the alternative pathway, while potentially bactericidal, was only slowly so. For these organisms at least, the alternative pathway could be expected to play only a minor role.

6. Proteus mirabilis

Leist-Welsh and Bjornson[49] have studied three clinical isolates of *Proteus mirabilis*. All three required immunoglobulin and complement for opsonization. Two of the three strains could utilize the alternative pathway in addition to the classical pathway.

7. Haemophilus influenzae

Haemophilus influenzae, primarily a pathogen of children, is usually associated with meningitis in the very young. The organism is readily isolated from normal upper respiratory passages. It may play some role in the exacerbations of chronic obstructive pulmonary disease. In vitro studies with activation of the complement system have given contradictory results. Using adult serum chelated with EGTA, Fine et al.[64] could not demonstrate activation of the alternative pathway by *H. influenzae* B. In contrast, Quinn and colleagues,[65] using the same species, demonstrated complement activation in normal and C4-deficient guinea pig serum. The amounts of complements consumed in both sera appeared quantitatively similar. In both studies,[64,65] no consumption of complement by the polyribosephosphate capsular material could be demonstrated.

Experiments in an infant rat model of *H. influenzae* B septicemia and meningitis[66] support the notion that complement plays an important role in defense against *H.*

influenzae. Treatment of these antibody-dependent rats with cobra venom factor led to an increased incidence of bacteremia, quantitatively greater bacteremia, and an increased mortality in a dose-dependent fashion. Cobra venom factor-treated rats had an increased incidence of meningitis.

Weller et al.[67] suggested clearance of *H. influenzae* B in mice was primarily by means of phagocytosis; there was no evidence for in vivo bacteriolysis. Clearance would be markedly enhanced by active or passive immunization. Newman et al.[68] also suggested that opsonization was more important than bacteriolysis.

Johnston et al.[69] showed that opsonization of *H. influenzae* B was dependent on or related to antibody titers, both hemagglutinating and bactericidal. In vitro opsonization, using sera obtained before and after vaccination, was enhanced by but not dependent on complement. This activation of complement apparently involved primarily the alternative pathway, since opsonization was only minimally inhibited by C2 deficiency but markedly inhibited by C3 deficiency.

VI. GRAM-NEGATIVE, ANAEROBIC BACTERIA

A. Bacteroides

Anaerobic Gram-negative bacteria are involved, probably only synergistically, in suppurative infections. *Bacteroides fragilis*, a normal bowel organism, is characteristically associated with abdominal or pelvic infections. However, the organism may also be isolated from purulent sinopulmonary infections. Casciato et al.[70] demonstrated serum bactericidal activity against a strain of *B. fragilis* in all sera tested. Among various strains, three of five stool isolates were serum sensitive, whereas only one of five blood isolates was serum sensitive. In subsequent studies,[71] these investigators found that 33% of strains of *B. fragilis* isolated from stool were sensitive to 10% serum, 50% were sensitive to 80% serum. In contrast, even in 80% serum, only 15% of isolates from sites of infection were killed. In the presence of polymorphonuclear leukocytes and serum, both serum-sensitive and serum-resistant strains were readily killed.[70] In contrast, Bjornson et al.[72] could demonstrate little or no serum bactericidal activity against *B. fragilis*. Their demonstration of effective phagocytosis in the presence of serum in aerobic or anaerobic environments suggested opsonophagocytosis as the major host defense mechanism. Subsequent investigations[73] established requirements for both immunoglobulin and complement, which functioned effectively even when the classical pathway was blocked.

B. fragilis lipopolysaccharides have been demonstrated to activate complement in vivo and in vitro.[74] The alternative pathway appeared to be involved. Likewise, lipopolysaccharides of *B. oralis* activated primarily the alternative pathway.[75] The general subject of lipopolysaccharide-complement interactions will be considered below.

B. Fusobacteria

These organisms, normal upper respiratory flora, have some role in purulent sinopulmonary infection. They are probably of limited pathogenicity. Bjornson et al.,[72] studying *Fusobacterium mortiferum*, demonstrated effective serum bactericidal activity in aerobic and anaerobic environments. They could demonstrate no enhanced killing in the presence of leukocytes, but the degree of killing by serum alone was so great as to obscure any further contribution from leukocytes. Hawley and Falkler[76] studied *F. polymorphum* (*F. nucleatum*), an organism isolated in large numbers from dental plaque and implicated in the etiology of periodontitis. Preparations of cell wall consumed guinea pig complement effectively, even in C4-deficient guinea pig serum. And lipopolysaccharides of *F. nucleatum* activated the complement system, though primarily through the classical pathway.[75]

VII. GRAM-NEGATIVE COCCI AND COCCOBACILLI (AEROBES)

A. Neisseria

Gram-negative cocci important in human disease include *Neisseria gonorrhoeae* and *N. meningitidis*. It will be remembered from the discussions of patients with terminal complement component deficiencies that there is a highly suggestive link between the complement system and host defenses against these major pathogens. Patients with deficiencies of terminal complement components (and thus without the ability to kill these organisms by serum alone) have an inordinate susceptibility to disseminated infection with these organisms.

1. Neisseria gonorrhoeae

Several studies have suggested that organisms isolated from patients with disseminated gonorrhea are generally serum resistant, whereas mucosal isolates from patients with uncomplicated gonorrhea are generally serum sensitive.[77,78] Even in the study in which they found 62% of local isolates to be serum resistant, Eisenstein and colleagues[79] noted an increase of serum resistance among strains isolated from disseminated gonococcal infection (97% serum resistant). When they studied only patients' sera against their own organisms, Brooks and Ingwer[80] could document only 60% of genital isolates to be serum sensitive; however, 80% were sensitive when tested against a variety of sera. On the other hand, Tramont et al.[81] found only 4 of 24 mucosal isolates from patients with uncomplicated gonorrhea to be serum sensitive and did not find a higher incidence of serum resistance among organisms isolated from patients with disseminated infection.

Eisenstein et al.[79] found penicillin sensitivity to correlate strongly with dissemination among strains of gonorrhea. Likewise, they found serum resistance to correlate with disseminated infection but to be an attribute independent of penicillin sensitivity.

Tramont et al.[81] further suggested that gonococcal infection does not regularly induce bactericidal antibody and that bactericidal activity of serum cannot explain uncomplicated infection. They suggested that the bactericidal antibody is probably not an important host defense against *N. gonorrhoeae*. Ingwer and colleagues[80,82] demonstrated that, in their patients with localized genital gonorrhea, serum was able to kill the organisms isolated or bactericidal antibodies eventually developed. This killing was heat-labile and inhibited by EGTA, indicating mediation primarily through the classical complement pathway. They demonstrated similar results with normal serum and laboratory strains of *N. gonorrhoeae*. The small amount of killing demonstrable in EGTA-chelated serum, i.e., via the alternative pathway, was slow and unlikely to be clinically relevant. They demonstrated that IgG enhanced killing by the alternative pathway. Bactericidal activity thus did not preclude genital infection.[80] However, serum from patients with disseminated infection could kill neither the organism isolated from their infection nor stock laboratory strains.[82] These results are consistent with the notion that serum bactericidal activity plays a role in defense against systemic infection. Likewise, Brooks et al.[77] demonstrated that patients with disseminated gonorrhea did not develop bactericidal antibody, though they did often develop hemagglutinating and complement-fixing antibody.

In an animal model of gonorrhea, mice congenitally deficient in C2 and C4 were immunized with piliated or nonpiliated *N. gonorrhoeae*.[83] Immunization with the piliated organisms enhanced resistance to subsequent subcutaneous challenge. Injection of complement, that is, normal mouse serum, further enhanced resistance. Immunization with the nonpiliated organism did not enhance resistance unless exogeneous complement was also added.

Schiller et al.[84] studied opsonization of avirulent *N. gonorrhoeae* by normal serum. Using a concentration of serum not bactericidal in the absence of leukocytes, they demonstrated a primary role for cross-reactive antibody and a lesser role for complement, primarily the alternative pathway.

2. Neisseria meningitidis

In a 1943 study of patients and carriers from Boston and Nova Scotia, Thomas and Dingle[85] demonstrated no correlation between mouse virulence and serum sensitivity. However, there are a number of studies to indicate activation of the complement system, for good or ill, in patients with meningococcal disease. In one patient with meningococcal meningitis and purpura, Johnson and Laurell[86] found low levels of C1q, C3, and properdin early in infection.

Hoffman and Edwards[87] demonstrated that, as a group, patients with meningococcal disease did not have particularly low levels of C3. However, the subgroup with antigenemia (20 patients) had lower mean values than the subgroup without antigenemia (37 patients). Furthermore, whereas 59% of antigenemic patients developed hypocomplementemia during the first 48 hr, only 19% of nonantigenemic patients did so. In 12 patients studied serially, C3 levels fell after antigen was no longer demonstrable; the decrease was assumed to represent formation of immune complexes. Greenwood et al.[88] studied 13 patients with acute meningococcemia. These patients had markedly depressed C3 levels (approximately one half normal). Patients with meningitis only (no demonstrable bacteremia) did not have depressed levels of C3. However, when they divided meningitis patients into those with and without antigenemia (as demonstrable by counterimmunoelectrophoresis), antigenemic patients again had C3 values statistically lower. Patients with arthritis did not have low levels of joint fluid complement.

VIII. GRAM-POSITIVE COCCI

A. *Staphylococcus epidermidis*

Often referred to as *Staphylococcus albus,* these normal skin flora are associated primarily with persistent, indolent infections of intravascular prosthetic devices, such as artificial heart valves or ventriculoatrial shunts (plastic tubes diverting fluid from the cerebral ventricles to the right atrium of the heart and used in the management of some kinds of hydrocephalus). The syndrome of chronic ventriculoatrial shunt infection, probably first described by Black and colleagues[89] in 1965, is characterized by fever, myalgias and arthralgias, nephritis, and sometimes nephrotic syndrome. Though the syndrome is not specific for *S. epidermidis,* at least 15 patients have been reported in whom *S. epidermidis* was the causative organism.[90-95] In general, the patients have had persistent bacteremia, hypocomplementemia (especially components of the classical pathway), and often cryoglobulinemia or rheumatoid factor activity. Renal biopsies have disclosed glomerular deposits of immunoglobulins, complement, and in a few cases, staphylococcal antigen.

A review of one considerable experience with ventricular shunts[96] disclosed 49 infections of ventriculoatrial shunts in 48 patients. Twenty-nine of these infections (approximately 60%) were due to *S. epidermidis.* Two of the 48 patients (4%) developed glomerulonephritis, but only after long-standing, inadequately treated in-

Table 3
IN VITRO COMPLEMENT ACTIVATION BY AND OPSONIZATION OF MISCELLANEOUS GRAM-POSITIVE BACTERIA

Organism	Activation		Opsonization			Ref.
	CCP^a	ACP^b	Normal serum	C^c	ACP^b	
Staphylococcus epidermidis	ND^d	ND	+	ND	ND	97
	ND	ND	+	+	+	25
	ND	ND	+	+	ND	46
	ND	ND	+	±	+	98
Listeria monocytogenes	+	+	ND	ND	ND	168
Erysipelothrix rhusiopathiae	+	+	ND	ND	ND	169
	ND	ND	+	+	ND	170
Corynabacterium parvum	+	ND	ND	ND	ND	174
	+	ND	ND	ND	ND	175
Propionibacteria	+	+	ND	ND	ND	176,177

[a] Classical complement pathway.
[b] Alternative complement pathway.
[c] Opsonization in normal serum is dependent on (+) or independent of (−) or only partially dependent on (±) complement.
[d] Not done.

fections; one of these two patients was infected with *S. epidermidis*. Thus, although this organism is a prominent cause of infected shunts, it seems to have no other unusual propensity to cause complement-mediated kidney disease.

In vitro studies (Table 3) have generally indicated that *S. epidermidis* is adequately opsonized by normal serum.[25,46,97,98] When studied, complement was required for this opsonization[25,46] and the alternative pathway seemed sufficient.[25]

B. *Staphylococcus aureus*

It can be seen from Table 4 that a number of investigators have examined both the interactions of strains of *Staphylococcus aureus* with the complement system and the complement requirements for opsonization. In general, most staphylococcal strains have activated complement and have required complement to be ingested and killed by phagocytes. There have been considerable variations in terms of the ability of the alternative pathway to function opsonically with the various strains. Indeed, differences have been seen even among experiments with the same strain.

Using various means to dissect the roles of immunoglobulin and the complement pathways in opsonization of *S. aureus* by normal serum, Verbrugh and colleagues[101] found that optimum phagocytosis — optimum in terms of rate — required immunoglobulin and an intact classical pathway. Their observations provided useful insight into the kinetics of opsonization. After 5 min in 5% serum, radiolabeled bacteria were maximally opsonized, i.e., further incubation with serum did not enhance subsequent uptake by leukocytes. By 5 min, C3 and immunoglobulin deposition on the bacterial surface was also maximal, as measured by quantitative immunofluorescence. (Interestingly, complement *consumption* could not be measured before 15 min.) At lower dilutions of serum, the same degree of opsonization could be achieved but only after 30 min. Serum in which immunoglobulin was deficient or

Table 4
IN VITRO COMPLEMENT ACTIVATION BY AND OPSONIZATION OF *STAPHYLOCOCCUS AUREUS*

	Activation		Opsonization			
Organism	CCP[a]	ACP[b]	Normal serum	C[c]	ACP[b]	Ref.
502 A	+	+	+	+	−	45
	ND[d]	ND	+	+	+	98
	ND	ND	+	+	ND	99
Cowan I	ND	ND	+	+	−	45
	ND	ND	+	+	+	98
Wood 46	ND	ND	+	+	−	45
	ND	ND	+	+	+	98
Unspecified	ND	ND	−	ND	ND	98
	ND	ND	+	+	ND	46
	+	+	+	+	+	100

[a] Classical complement pathway.
[b] Alternative complement pathway.
[c] Opsonization by normal serum is dependent on (+) or independent of (−) complement.
[d] Not done.

in which the classical pathway had been blocked (C1s or C2 deficiency or MgEGTA chelation) supported optimum opsonization only after 30 to 60 min incubation with *S. aureus*. In related experiments, Murphey et al.[102] demonstrated opsonization of *S. aureus* by nonimmune serum for alveolar macrophage phagocytosis. This opsonization was complement dependent and could utilize the alternative pathway though opsonization was not optimum. When immune serum was the opsonic source, a role for complement was demonstrable only at low antibody concentrations (0.3% serum). These results[102] and those of Verbrugh et al.[101] emphasize the complexity of opsonization and the need to evaluate results in the light of methodology.

The effect of encapsulation on staphylococcal opsonization has been studied (Table 5).[103-105] With the strain M, encapsulation completely blocked opsonization by normal serum. Strain Smith, though it could be opsonized in encapsulated form by normal serum when only the alternative pathway was functional, was not so well opsonized as in the presence of an intact complement system. Peterson and colleagues[103,104] concluded that encapsulation blocked or inhibited opsonization, especially by the alternative pathway. Koenig et al.[105] demonstrated that an encapsulated staphylococcus could be opsonized in an isolated perfused liver system by normal serum only if complement were present, whereas the unencapsulated variant could be opsonized by normal serum even in the absence of complement.

The staphylococcal capsule did not, in the studies of Peterson et al.,[104] inhibit complement activation through either pathway in terms of rate or quantity. Thus, activation of complement was not necessarily synonymous with opsonization, a conclusion also reached in studies of pneumococci.[106] Wilkinson et al.[107,108] found that the capsule did not inhibit activation or deposition of complement on bacterial surfaces, but rather it interposed a barrier between complement on the bacterial cell wall and receptors on phagocytes (Figure 1). Thus, complement could be envisioned

Table 5
EFFECT OF CAPSULE ON OPSONIZATION OF *STAPHYLOCOCCUS AUREUS*

Strain	Capsule	Opsonization Normal serum	C[a]	ACP[b]	Ref.
Smith	+	+	+	+	103,104
	−	+	+	+	103,104
M	+	−	−	−	103,104
	−	+	+	+	103,104
Unspecified	+	+	+	ND[c]	105
	−	+	−	ND	105

[a] Opsonization by normal serum dependent on (+) or independent of (−) complement.
[b] Alternative complement pathway.
[c] Not done.

as buried deep in the capsule and inaccessable to opsonic receptors.[104,107,108] Anticapsular antibody opsonized by reacting with the true outer surface of the organism, where antibody and complement could be recognized by phagocyte receptors.[108]

However, quantitative immunofluorescence studies suggested that another mechanism might explain the antiphagocytic properties of *S. aureus* capsules.[109] Encapsulated organisms (*S. aureus* as well as *E. coli, K. pneumoniae,* and *Salmonella typhi*) fixed less C3 to their surfaces than did nonencapsulated bacteria when incubated in 20% nonimmune serum for up to 30 min. There was a strong correlation between the amount of C3 bound to the organism and subsequent phagocytic indexes. These data do not necessarily contradict those of Wilkinson et al.,[108] who did not use a quantitative assay of C3 fixation and who also incubated bacteria in 100% serum. Thus, capsules may inhibit opsonization in two ways: less complement could be fixed to the bacterial surface and, in addition, that fixed complement could be obscured from phagocytes. It seems likely that the major "antiphagocytic" activity of capsules of *Staphylococcus aureus* and probably other bacteria lies in some such inhibition of effective opsonization.

There is considerable controversy as to the underlying cell wall structure of staphylococci responsible for activation of the alternative pathway, probably either peptidoglycan or teichoic acids. The latter have been strongly implicated as alternative pathway activators by Saulsbury and Winkelstein.[110] They used cell-wall-deficient, L-phase variants of staphylococci. Such variants lack peptidoglycan but have considerable cell membrane-associated teichoic acids. These L-phase variants consumed C3 in C4-deficient guinea pig serum at a rate and quantity equivalent to parent strains and were killed by phagocytes in the presence of C4-deficient guinea pig serum normally. This opsonic activity was heat labile and dependent on divalent cations. Isolated L-phase membranes also consumed C3.

In contrast, other investigators have suggested that it is the peptidoglycan of *S. aureus* which reacts with serum opsonins. Shayegani et al.[111] demonstrated that peptidoglycan from *S. aureus* 3528 could absorb opsonins from normal serum. Digestion of the peptidoglycan with lysozyme destroyed this ability. Teichoic acids and

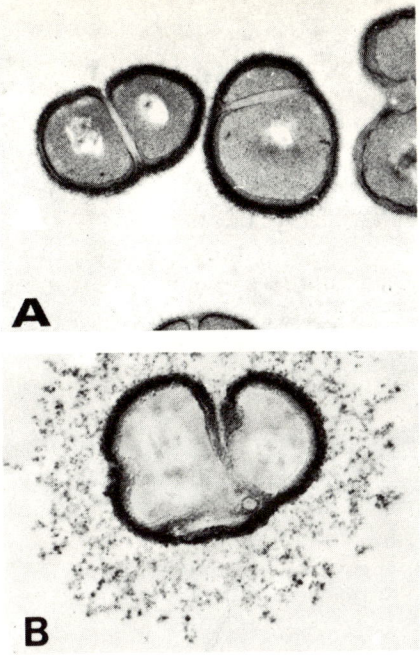

FIGURE 1. Deposition of C3 on cell wall beneath capsule. Encapsulated *Staphylococcus aureus* M was incubated with normal human serum at 37°C for 15 min, then at 4°C for 60 min with goat antiserum to human C3 conjugated with horseradish peroxidase. (A) Organisms were then incubated at 0°C with rabbit antibody (IgG) to *S. aureus* M to preserve the capsule, or (B) with the rabbit antibody to *S. aureus* M plus horseradish peroxidase-labeled goat antiserum to rabbit IgG to preserve and demonstrate the capsule. Bacteria were then examined by electron microscopy.

Presence of the peroxidase-labeled anti-C3 at the cell wall in both pictures indicates that C3 was deposited at that site during the incubation in normal serum. Figure 1B further demonstrates that the C3 was deep to the peroxidase-stained capsule and presumably inaccessible to phagocyte receptors. (From Wilkinson, B. J., Peterson, P. K., and Quie, P. G., *Infect. Immun.*, 23, 502, 1979. With permission.)

protein A could not absorb the opsonin. Peterson, Wilkinson, and colleagues[107,112] have also presented data to implicate peptidoglycan in the opsonization of *S. aureus*. Their studies were performed with *S. aureus* strain H, the rough streptomycin-resistant mutant of strain H, and the teichoic acid-deficient strain 52A5. Furthermore, they purified peptidoglycan and the ribitol teichoic acids from staphylococci. Using radiolabeled particles, they demonstrated that phagocyte uptake of intact *S. aureus* H, the peptidoglycan from that organism, or the complete cell walls (containing both peptidoglycan and teichoic acids) were similar in terms of the percentage of radioactivity that became cell associated and the percentages of serum that supported phagocytosis. They demonstrated some degree of phagocytosis even in heat-inactivated serum; similar levels of phagocytosis could be obtained using purified IgG. The rough mutant of *S. aureus* H and the teichoic acid-deficient strain 52A5 were phagocytosed to the same degree in normal and in heat-inactivated serum. Phagocytosis of bacteria and peptidoglycan occurred in serum that had been chelated with MgEGTA and in C2-deficient serum. Peptidoglycan could activate the alternative pathway; this activation required an absorbable noncomplement factor. In studies of the heavily encapsulated *S. aureus* M, they found that the capsule served to obscure opsonization of the cell wall structures. They suggested that peptidoglycan was the substance that activated the complement system and was the recognition part of the bacterial particle. The polysaccharide capsule served to obscure either the recognition unit or the recognition by the phagocyte of the deposited complement, as noted above.

In studies of serum chemotactic factor generation, Schmeling and colleagues[113] found that purified peptidoglycan, while effective, was less so than purified cell wall preparations, which also contained teichoic acids. They suggested a role for both components in chemotaxigenesis. The same conclusion may also apply to opsonization as well.

Staphylococci produce a curious protein that is excreted and also found on the surface of the bacterium. This protein is called staphylococcal protein A (SPA). Sjöquist and Stalenheim[114] demonstrated that this protein consumed complement. SPA can react with and so alter the Fc fragment of IgG as to activate complement.[115] These observations raised the intriguing possibility that this protein, which is shed from staphylococci during growth phases, may actually divert the complement opsonins. Forsgren and Quie[116] found that the opsonization of *S. aureus* could be impaired by incubation of serum with SPA; they suggested that this effect was in fact a defense mechanism of the bacteria. Furthermore, Peterson et al.[117] demonstrated that strains of *S. aureus* lacking ability to make SPA could be phagocytosed at a faster rate than could the SPA-rich strains.

Easmon and Glynn[118] suggested that C3 was important in defense against staphylococcal infections. In their studies, subcutaneous and intraperitoneal injections of staphylococci produced more serious systemic and local disease in mice depleted of C3. C5-deficient mice also appeared to be more susceptible to staphylococcal infections. On the other hand, Gross et al.[27] could not show that complement played a role in clearance of aerosols of *S. aureus* in mice. In a different model (subcutaneous staphylococcal infections),[119] C5 deficiency appeared to predispose animals to larger numbers of bacteria in skin lesions but otherwise to little excessive disease.

C. Streptococci
1. Streptococci other than Streptococcus pneumoniae

Streptococci have been divided by Lancefield into serological groups; this grouping has considerable pathogenic importance. There are now 80 or more groups, identified alphabetically. Some of the more commonly encountered groups and species of streptococci within these groups are listed in Table 6, which is derived from the extensive review by Duma et al.[120] Organisms that may not be grouped according to the Lancefield scheme are called *Streptococcus viridans* or, more properly, viridans streptococci. *S. pyogenes,* the major pathogen of group A, produces colonies on blood agar that are usually surrounded by a clear zone attributable to hemolysins produced by the streptococci. These group A, beta-hemolytic streptococci are major causes of pharyngitis and skin and wound infections. Group B streptococci are major causes of genital infections and infections of neonates. Group C streptococci are involved, in general, in skin and wound infections. Among group D, *S. faecalis* is a cause of urinary tract infections and endocarditis. Other group D organisms, *S. bovis* and *S. equinus*, also cause endocarditis. The viridans streptococci may produce many of the same clinical infections as groupable streptococci can produce; but classically, viridans streptococci are associated with infective endocarditis.[120]

In 1936 Keefer and Spink[121,122] reviewed an 11-year experience at the Boston City Hospital with 1400 cases of hemolytic streptococcal infections. They noted marked variability throughout infection in the ability of patients' blood to kill streptococci. This streptococcidal power showed no particular pattern and did not correlate with hemolytic complement levels. The streptococcidal power of defibrinated blood was

Table 6
CLASSIFICATION OF STREPTOCOCCI[a]

Lancefield groups
  A: *S. pyogenes*
  B: *S. agalactiae*
  C: *S. equi, S. zooepidermicus, S. equisimilis, S. dysgalactiae*
  D: *S. faecalis* (enterococcus), *S. faecium, S. bovis, S. equinus*
  E: *S. infrequens*
  F: *S. minutus, S. anginosus*
  G: *S. canis*
  H: *S. sanguis*
  K: *S. salivarius*
  N: *S. lactis, S. cremoris*
Nongroupable
  Viridans streptococci
S. pneumoniae

[a] Adapted in part from Duma et al.[120]

a measure of the cidal ability of serum and polymorphonuclear leukocytes and thus measured both opsonins and phagocytes. No other paper dealing specifically with changes in complement activity or opsonic activity in streptococcal infections (excluding glomerulonephritides) was identified in this review.

Table 7 summarizes information on in vitro ability of streptococci to activate complement by various pathways and to be opsonized in normal serum. Among these relatively few studies, *S. pyogenes* has received most attention, with conflicting results. It may be pointed out that the virulent streptococcus studied by Greenblatt et al.[123] did not activate the alternative pathway, whereas the nonvirulent organism studied by Sakai et al.[125] did activate the alternative pathway. The number of strains studied, however, has been too small to draw conclusions.

Group B streptococci have generally appeared unable to be opsonized by normal serum and have required specific opsonins (i.e, specific antibody) to be ingested and killed by phagocytes.[126,127] This very specific requirement seems to correlate with the documented propensity of this organism to cause overwhelming infection in neonates.

A number of studies have been done to identify the streptococcal cell wall constituent responsible for activation of complement, especially by the alternative pathway. Resnick and Becker[128] demonstrated that cell walls of groups A, B, C, D, and G streptococci were capable of activating complement in serum chelated with EGTA, thus through the alternative pathway. Greenblatt et al.,[123] studying primarily group A streptococci, suggested that the responsible structure was the peptidoglycan, not the cell membrane. (Intact cell walls consumed only limited amounts of complement.) In studies with *Streptococcus mutans*, Inai et al.[129] suggested also that the reactivity with complement was in the glucan structure and the associated side chains. On the other hand, Saulsbury and Winkelstein[110] studying *S. faecalis*, Silvestri et al.[130] studying *S. mutans*, and Tauber et al.[131] studying cell membranes from groups A, C, D, and G all suggested that complement reactivity was associated with cell membranes and was likely to be due to teichoic acids. As with staphylococci, the question of which cell wall constituents activate the alternative pathway remains unresolved.

However, it has been demonstrated that whatever the structure which activates the alternative pathway, peptidoglycan or teichoic acids, the M protein of strepto-

Table 7
IN VITRO COMPLEMENT ACTIVATION BY AND OPSONIZATION OF STREPTOCOCCI

Organism	Activation		Opsonization			Ref.
	CCP[a]	ACP[b]	Normal serum	C[c]	ACP[b]	
A	ND[d]	ND	−	ND	ND	97
	ND	—	ND	ND	ND	123
	ND	ND	+	−	ND	124
	ND	+[e]	ND	ND	ND	125
B	+	+	−	ND	ND	126
	ND	ND	+	ND	ND	127
D	ND	ND	+	+	+	25
Viridans	ND	ND	+	+	+	25

[a] Classical complement pathway.
[b] Alternative complement pathway.
[c] Opsonization by normal serum dependent on (+) or independent of (−) complement.
[d] Not done.
[e] "Nearly avirulent" strain.

cocci serves to block opsonization via that activation.[132] M protein, a streptococcal product long associated with virulence, is found on the fimbriae of healthy bacteria in association with lipoteichoic acid. A strain of *S. pyogenes* that did not produce M protein was readily ingested by leukocytes in normal serum. Opsonization was not inhibited by MgEGTA-chelation of normal serum and was adequate in C2- or immunoglobulin-deficient sera. These data indicated that the M-protein-deficient streptococcus activated the alternative pathway and was thereby opsonized. In contrast, a variant capable of producing large amounts of M protein was poorly ingested, if at all, by polymorphonuclear leukocytes in the presence of normal serum; what phagocytosis occurred was not inhibited by MgEGTA. Digestion of M protein with trypsin or pepsin enhanced opsonization in normal human serum. These authors[132] suggested that M protein obscured the structure, perhaps peptidoglycan, that activated the alternative pathway and led to opsonization. Alternatively, the M protein may have interfered with interaction of deposited complement components and polymorphonuclear leukocytes. Bisno[133] has presented evidence for the former hypothesis. M-protein-containing streptococci consumed considerably less complement in MgEGTA-chelated serum than did M-protein-deficient variants. Antibodies to the M protein are major opsonins in immune serum and thus are presumably able to overcome the phagocytosis-inhibiting effect of M protein.

It has also been suggested[134,135] that streptococci produce a protein, analogous to staphylococcal protein A, that may also react with immunoglobulins. It is not known whether that interaction leads to complement consumption.

The glomerulonephritis following group A streptococcal (*S. pyogenes*) infections appears to be mediated by immune complexes and complement.[136,137] Hypocomplementemia early in the disease is common.[138] A similar glomerulopathy may occur as a complication of endocarditis caused by other streptococci, such as *S. faecalis*[139] and viridans streptococci.[140,141]

2. *Streptococcus pneumoniae*

Streptococcus pneumoniae (formerly *Diplococcus pneumoniae*) is a Gram-positive bacterium commonly found in the upper airway of normal individuals. Organisms of this species are grouped on the basis of serological reactions that reflect variability of the capsules. The capsules are generally polysaccharides ranging from very simple polymers of glucose to very complex polysaccharides. There are well over 80 serotypes of *S. pneumoniae*. Austrian and Gold[13] demonstrated in adults a hierarchy of virulence among pneumococcal strains; in general, lower numbered serotypes (e.g., 1, 3, 4, etc.) were more virulent than higher numbered serotypes, as reflected by more frequent isolation of the former from blood during serious pneumococcal pneumonia. Lund[142] and Mufson et al.[143] have confirmed these observations. Correlation of virulence with lower numbered serotypes would not be, in and of itself, surprising, since it seems logical that the more pathogenic a serotype, the more commonly it would be encountered and thus the more likely it would be to be numbered first.

An important relationship between pneumococcal infections and the complement system has been recognized for some time, and interactions of pneumococci and complement are perhaps better studied than those of any other Gram-positive organism. Dick[144] in 1912 measured, in serial blood samples from patients with pneumococcal pneumonia, the ability of blood to "digest" pneumococci, as reflected by decrease in optical rotation. Blood (including serum products and leukocytes) obtained following the crisis had enhanced digestive ability, which he correlated with increased complement levels. However, the increased digestibility appeared specific for pneumococci.

Rutstein and Walker[145] studied 75 patients in 1942 with pneumococcal pneumonia. Diagnoses were based on roentgenographic evidence of pneumonia and demonstration of pneumococci in the sputum; these criteria would not be considered adequate today, but most of their patients probably did have pneumococcal pneumonia. Approximately 17% of their patients had low complement levels at the time of admission to the hospital; all complement levels returned to normal after recovery. Eight of the ten patients who died had normal complement values at admission, two had low values (not a significantly higher incidence than in the general study group). Six of seven patients who had multiple complement determination and who eventually died developed lower values close to death. Bacteremia did not correlate with low complement values. Capsular type may have correlated with low complement values; all patients with low complement values were infected with types 1, 3, or 7.

In 22 patients with pneumococcal infections studied more recently, 21 of whom had pneumonia, complement levels were assayed at admission and at recovery.[146] Classical pathway proteins (C1q and C4) were normal, whereas alternative pathway proteins (factor B, C3) were low at the time of admission. The authors concluded that either the alternative pathway had been activated during acute infection or that a prior low level of alternative pathway function (perhaps from alcoholism) predisposed to infection. And Coonrod and Rylko-Bauer[147] also demonstrated decreased levels of C3 and properdin in patients with acute pneumococcal pneumonia. They could not demonstrate decreased levels of factor B. Convalescing patients had high levels of C1q, C3, C5, and factor B. C3 or factor B breakdown products were not demonstrable. However, alternative pathway function of acute sera (the amount of complement consumed by a given amount of zymosan) was significantly depressed. Bacteremic patients had lower values of C3 than nonbacteremic patients, who did not differ from normal. Complement levels could not be related to serotypes causing infection.

In experimental animals, there is also evidence for the importance of the complement system in pneumonia and pneumococcal disease. Gross et al.[27] demonstrated that pulmonary clearance of aerosolized *S. pneumoniae* (type unknown) required complement. Guckian et al.[148] studied a rabbit model of pneumococcal pneumonia (as distinguished from aerosolization). Type 25 pneumococci were not virulent unless animals had been depleted of complement by prior treatment with cobra venom factor, in which case animals developed overwhelming pneumonia. In contrast, Young et al.[149] found that the volume of lung involved in experimental pneumococcal pneumonia (in rats) was diminished by prior treatment with cobra venom factor. Robertson et al.[150,151] studied experimental pneumococcal (serotype 1) pneumonia in the dog. Those dogs with an eventually fatal pneumonia almost uniformly developed bacteremia. Approximately half of the animals developed decreased pneumococcidal power of blood, in their studies a reflection of serum opsonic activity. There was rough correlation between numbers of organisms in the blood and loss of opsonic activity.

Ward and Enders[152] studied the opsonins in normal serum and immune serum for pneumococci (serotypes 1, 2, and 3). They suggested that the antibody specific for the carbohydrate moiety of the organism was responsible for most of the opsonic activity of normal human serum. They felt that complement contributed relatively little to the opsonic activity and they suggested that complement alone could not opsonize any of these bacteria. Jeter et al.,[153] on the other hand, inhibited phagocytosis of type 1 pneumococci by human neutrophils in immune serum with antiserum to complement. Their data however suggested the effect was on the leukocytes, since preincubation of leukocytes with the antiserum blocked subsequent phagocytosis. One can guess that they were actually measuring the effect of blocking the Fc receptor.

Among in vitro studies, it was reported that one of the more virulent serotypes, type 1, did not activate the alternative pathway in adult human serum.[154] This inability correlated with a specific opsonic requirement for immune serum in vivo and in vitro.[148,152] Types 3, 4, and 8 activated the alternative pathway, but absorption of the serum at 0°C with the serotype in question abrogated this ability.[154] In contrast, types 7, 12, 14, and 25 activated the alternative pathway in serum absorbed at 0°C with organisms of the same serotype. This relative nonspecificity of complement activation of these higher serotypes was suggested to be related to their lesser virulence.[154] Studies relating to this possibility are summarized in Table 8. As noted earlier, complement activation has not necessarily been a good reflection of opsonization.[44,106] Furthermore, many of these studies did not adequately evaluate kinetics of opsonization. With pneumococci, Giebink et al.[155] have emphasized that opsonization via the alternative pathway is generally slower and may require higher concentrations of serum. Nevertheless, the data do corroborate the variability among serotypes in activation of complement pathways and in opsonic requirements; such variability may be one correlate of pathogenicity.

There is agreement among investigators who have studied the question that the polysaccharides of the pneumococcal capsule do not activate complement by either pathway in the absence of highly specific antibody.[154,156,159,162] Neither do the capsules appear to block complement activation.[154,159] Dhingra et al.[162] suggested that the peptidoglycan of the cell wall activated complement. Winkelstein and Tomasz[163] found that purified cell walls of pneumococci were most active in complement fixation. They could show little activity in the cell membranes. Subsequent studies

Table 8
IN VITRO COMPLEMENT ACTIVATION BY AND OPSONIZATION OF PNEUMOCOCCI

Organism	Activation CCP[a]	ACP[b]	Opsonization Normal Serum	C[c]	ACP[b]	Ref.
1	ND[d]	−	ND	ND	ND	154
	+	±	ND	ND	ND	156
	ND	ND	+	+	+(?)	157
	+	+	ND	ND	ND	158
	ND	+	ND	ND	ND	159
2	+	+	ND	ND	ND	159
3	ND	+	+	+	+	106,154
	+	+	ND	ND	ND	156
	+	+	ND	ND	ND	158
	+	+	ND	ND	ND	159
	ND	ND	+	+	ND	160
4	+	+	ND	ND	ND	156
	ND	+	ND	ND	ND	154
	ND	±	ND	ND	ND	158
	+	+	ND	ND	ND	159
6	+	+	ND	ND	ND	156
	±	±	ND	ND	ND	158
	ND	ND	+	+	−	155
7	ND	+	ND	ND	ND	154
	+	±	ND	ND	ND	156
	±	±	ND	ND	ND	158
8	ND	+	ND	ND	ND	154
	±	±	ND	ND	ND	158
12	ND	+	ND	ND	ND	154
	±	±	ND	ND	ND	158
14	ND	+	ND	ND	ND	154
	+	+	ND	ND	ND	156
	+	+	ND	ND	ND	158
	+	+	ND	ND	ND	159
18	ND	ND	+	+	+	155
19	+	+	ND	ND	ND	158
	+	+	ND	ND	ND	159
23	ND	ND	+	+	−	155
25	+	+	+	+	+	48,106, 148,154, 159, 161
	+	±	ND	ND	ND	156
	ND	ND	+	+	+	25
	ND	ND	+	+	+(?)	157
	+	+	ND	ND	ND	158
	ND	ND	+	+	−	155

[a] Classical complement pathway.
[b] Alternative complement pathway.
[c] Opsonization by normal serum dependent on (+) or independent of (−) complement.
[d] Not done.

by these authors[164] suggested that the complement-activating fractions of the cell wall were teichoic acids. As with other Gram-positive bacteria, that question may be considered to be still relatively unsettled.

Pneumococci have also been suggested to consume complement through interaction with Fc fragments of immunoglobulin.[158,165] This claim has been challenged.[166]

Pneumococcal infections may be associated with secondary, complement-mediated glomerulonephritis. In one case,[167] pneumonia caused by *S. pneumonia* serotype 14 was complicated by alternative-pathway-mediated complement activation. Serum levels of C3 were low, whereas C4 levels were generally normal; the patient's serum converted C3 in vitro and this conversion was not inhibited by EGTA. Finally, renal glomeruli had granular deposits of C3, properdin, and type 14 antigen.

IX. GRAM-POSITIVE, ASPOROGENOUS ROD-SHAPED BACTERIA

A. *Listeria monocytogenes*

These organisms are uncommonly encountered as pathogens and then primarily in immunologically compromised patients. The only study[168] identified in this review indicated that the organism activated both complement pathways (Table 3). In addition, the organism has been isolated from at least one patient with ventriculoatrial shunt infection and hypocomplementemic nephritis.[95]

B. *Erysipelothrix rhusiopathiae*

Cellulitis and lymphangitis occur from traumatic occupational exposure to this organism, characteristically among fishermen. Strains have generally been readily reactive with complement, including the alternative pathway (Table 3).[169,170]

X. ACTINOMYCETES AND RELATED ORGANISMS

A. Corynebacteria

With the exception of *Corynebacterium diphtheriae*, the causative agent of diphtheria, the corynebacteria are common saprophytic skin bacteria of little consequence; their primary bad habit is a tendency to contaminate otherwise negative blood cultures. Caren and Rosenberg[171] studied the role of complement in resistance to infection with *C. kutscheri* in mice. They compared complement-deficient mice (C5-deficient) with complement-sufficient mice. Though the complement-deficient mice died more quickly and had a greater mortality at a given organism dose, they were still relatively resistant compared to strains of mice innately more sensitive. Nor did the presence of complement retard development of endogenous infection in steroid-treated mice. Radiolabeled organisms were not distributed differently in the absence of C5.

Cases have been reported of corynebacterial ventriculoatrial shunt infections associated with hypocomplementemic immune-complex glomerulonephritis. Strains involved were *C. bovis*[172] and diphtheroids.[173]

Suspensions of heat-killed *C. parvum* have been used as nonspecific immune stimulants in therapy of cancer. It has been suggested that the inflammation produced by injection of *C. parvum* may be mediated to some degree by complement, since the organisms can activate the classical pathway.[174,175]

B. Propionibacteria

These normal skin organisms may play some role in acne. They can activate both complement pathways (Table 3).[176,177] One ventriculoatrial shunt infection with complement-mediated nephritis has been associated with *Propionibacterium acnes*.[178]

C. Mycobacteria

Mycobacteria include the agents causing typical (*Mycobacterium tuberculosis*) and atypical (*M. cheilonei, M. intracellulare, M. kansasii,* etc.) tuberculosis and leprosy (*M. leprae*). There has been little evidence for any important involvement of the complement system in host defenses against or disease produced by these organisms, with the possible exception of the syndrome of erythema nodosum leprosorum (see below). However, the subject has been exposed to only limited investigations.

1. Mycobacterium tuberculosis

The purified protein derivative (PPD) antigen used for skin testing activated complement in rabbit, human, and guinea pig serum, primarily by the classical pathway though no responsible antibody was demonstrated in nonimmune serum. In rabbit plasma, this complement activation resulted in platelet aggregation.[179]

2. Mycobacterium leprae

Involvement of complement has been studied primarily by measuring complement activity in patients with various forms of leprosy. Two clinical patterns may occur in leprosy, tuberculoid and lepromatous. The former is associated with localized granulomatous lesions of skin and peripheral nerves; host defenses seem relatively strong and lesions are controlled, though often with significant nerve damage. Tuberculoid leprosy is relatively benign. Lepromatous leprosy, associated with apparently inadequate host defenses, is characterized by diffuse bacillary involvement of skin and peripheral nerves and a malignant progressive course. The syndrome of erythema nodosum leprosorum, usually associated with initiation of treatment in a patient with lepromatous leprosy, is characterized by cutaneous vasculitis and fever. Histologic examination of the skin lesions discloses perivascular accumulation of polymorphonuclear leukocytes, fibrinoid necrosis of vessel walls, and endothelial swelling. Wemambu et al.[180] likened this picture to the Arthus phenomenon and demonstrated granular deposits of immunoglobulins (primarily IgG) and complement in the vessel walls in areas corresponding to leukocyte accumulations. In some biopsies, they could identify mycobacterial antigens. They could not demonstrate diminished serum levels of C3 in these patients.

De Azevedo and de Melo[181] demonstrated low complement levels in sera from patients with erythema nodosum leprosorum but not patients with tuberculoid or uncomplicated lepromatous leprosy. These results could not be confirmed in a small number of patients by Saitz et al.[182] Petchclai et al.[183] demonstrated anticomplementary activity in serum from lepromatous patients with a history of erythema nodosum leprosorum and decline in the complement activity of such serum with prolonged storage. This observation may explain differences among results by various investigators. Petchclai et al.[183] found variable minor increases and decreases of complement components in all leprosy patients studied; no clear patterns emerged. More recently Srivastava et al.[184] found normal C4 levels and decreased C3 levels in lepromatous and tuberculoid leprosy; they could not demonstrate any relationship of C3 phenotoypes and propensity to develop leprosy.[185]

D. Micromonosporaceae

Inhalation of moldy hay dust is the precipitating event in the disease called farmers' lung, an occupationally-related hypersensitivity in people who work around hay. The disease is characterized by dyspnea, wheezing, and fever on exposure to moldy hay. Many patients have precipitating antibodies to certain actinomycetes (primarily

Micropolyspora faeni and *Thermoactinomyces vulgaris*); these antibodies are thought to combine with inhaled antigens in the lung, following which complement activation and leukocyte accumulation produce the inflammation that interferes with oxygen exchange.[186]

However, some patients lack demonstrable antibodies and it has been suggested that these actinomycetes may be able to activate complement through the alternative pathway in the nonimmune patient.[186] Edwards and colleagues[187] demonstrated electrophoretic conversion of factor B when fractions of moldy hay or hay dust were incubated with normal human serum. Later, Edwards[186] showed complement consumption in MgEGTA-chelated serum (i.e., by the alternative pathway) by *M. faeni*, *T. vulgaris*, and, to a lesser extent, hay dust. *M. faeni*, the most potent activator of the alternative pathway, consumed complement as efficiently as did zymosan. Marx and Flaherty[188] demonstrated activation of the classical pathway by extracts of *M. faeni* and *T. vulgaris* in the presence of antibody, of the alternative pathway in the absence of antibody.

Burrell and McCullough[189] let nonimmune animals inhale *M. faeni* or *T. vulgaris*. Animals developed a variable and mild decrease in arterial oxygen concentrations and concomitantly a 21 to 23% decrease in hemolytic complement.

XI. THE RICKETTSIAS

A. *Rickettsia rickettsii*

Rickettsia rickettsii is the causative agent of Rocky Mountain spotted fever, a disease rarely seen in the Rocky Mountains. The characteristic pathology of the disease is a widespread vasculitis;[190] the classical rash reflects this vasculitis. Thrombocytopenia is quite common during the illness, and disseminated intravascular coagulopathy may supervene. Graybill and colleagues[191] demonstrated decreases in both whole hemolytic complement (CH50) and C3 (radial immunodiffusion) in association with thrombocytopenia and coagulation disturbances. They suggested that complement activation (e.g., by immune complexes) might initiate the coagulopathy or perhaps vice versa.

Guinea pigs experimentally infected with *R. rickettsii* developed a vasculitis that occurred prior to any detectable antibody and directly correlated with degree of rickettsemia. Complement values (CH50) were modestly depressed from day 2 of infection through day 10.[192] Vascular lesions studied 8 days after a similar experimental infection had C3 deposited in the vascular walls.[193]

In contrast, Mosher et al.[194] documented increased C2 and C3 levels in rhesus monkeys infected with *R. rickettsii*. One animal out of 27 developed a slight fall in complement on day 5 of infection in association with the appearance of antibody; that animal alone later developed peripheral gangrene. These investigators suggested two pathophysiological phases of Rocky Mountain spotted fever: an early "toxemic" phase produced directly by rickettsiae or rickettsial products without participation of complement and a late phase during which vascular necrosis was a feature and in which immune complexes and complement might be important. Supporting this hypothesis were serial evaluations of a patient with Rocky Mountain spotted fever.[195] The disease was suspected and observations begun unusually early in the course because he worked in a laboratory in which *R. rickettsii* was being studied. Rash, thrombocytopenia, and disseminated intravascular coagulopathy developed without any decreases in hemolytic complement or C3 and without appearance in plasma of C3 breakdown products.

XII. THE MYCOPLASMAS

These organisms lack a cell wall and exist in nature protected only by the cell membrane. They were first isolated from cattle with a distinctive respiratory illness called pleuropneumonia. From their relationships to this organism came the designation of subsequently identified mycoplasmas as pleuropneumonia-like organisms (PPLO).[196] This older term is no longer preferred. The analogy of mycoplasmas to cell-wall deficient variants of other bacteria adds interest to study of their interactions with complement and other serum proteins. Muschel and Jackson[197] demonstrated the serum-resistant bacteria became, on conversion to cell-wall deficient protoplasts, susceptible to the lytic effect of complement.

A. *Mycoplasma pneumoniae*

This organism causes an acute self-limited respiratory disease in man (primary atypical pneumonia, mycoplasmal pneumonia) and is, of all human mycoplasmas, the most virulent. The facts that mycoplasmal pneumonia is predominatly a disease of young people and occurs sometimes in epidemics point to the importance of an immune response to the organism. However, other features of the illness, such as the greater severity associated with infections in older patients, have suggested to some that immunological responses may produce some of the pathology.[198]

Eaton and colleagues[199] demonstrated cytopathic changes induced by the organism in tissue culture. Immune serum could block the cytopathology; a heat-labile factor was required. Growth on artificial media was also inhibited by immune serum and a heat-labile serum factor.[200,201] In at least one study, high concentrations (1:10 dilution) of nonimmune serum also inhibited growth.[200]

Mycoplasmacidal activity (as distinct from growth inhibition) of immune serum appeared to follow the same kinetics and display the same characteristics as complement-mediated hemolysis: a required factor was heat labile, blocked by EDTA, sensitive to salt concentrations, absorbable with immune complexes, and inactive at 4°C.[202,203] Electron microscopy disclosed membrane "ghosts" with lesions resembling those induced by complement.[204,205]

Effects of serum on mycoplasma include "rounding" (a morphological change associated with delayed multiplication and decreased adherence to host cells) and actual mycoplasmal death. Bredt and Bitter-Suermann[206] demonstrated that nonimmune guinea pig and rabbit sera could induce rounding; "rounding activity" was destroyed by heat (56°C, 30 min) but was not inhibited (or was only slightly inhibited) by MgEGTA. Rounding occurred in C4-deficient guinea pig serum and C6-deficient rabbit serum. C4-deficient guinea pig serum depleted of factor D could not induce rounding; rounding activity was restored by factor D. Rounded mycoplasma were opsonized. In contrast, mycoplasmacidal activity (of a 1:10 serum dilution) was abolished by heat inactivation, C4 or C6 deficiency, and MgEGTA. The authors concluded that complement-mediated killing required an intact classical pathway, but that alternative pathway activation could damage (round) and opsonize mycoplasmas. Bredt[207] subsequently demonstrated that opsonization was not optimum in C4-deficient guinea pig serum (as indicated by delayed onset of phagocytosis).

In contrast to earlier observations that killing of mycoplasma generally required immune serum (except perhaps by very high concentrations of nonimmune serum),[199-202,204] Bredt and colleagues[208] have recently shown both rounding and killing in antibody-depleted serum diluted even to 1:80. They suggested that killing required in such circumstances a direct interaction between the mycoplasmas and

C1 with subsequent activation of the classical pathway. Rounding also appeared to be at least enhanced by the reaction with C1.

Loos and Brunner[209] noted early and dramatic increases in complement components in guinea pig bronchial secretions following experimental *M. pneumoniae* infection. They suggested these increases could represent a nonspecific but primary host defense.

B. *Mycoplasma hominis*

Mycoplasmas may be isolated from the human genital tract, where they probably cause urethritis and perhaps other syndromes. Edward and Fitzgerald[210] studied serum effects on "pleuropneumonia-like organisms" from the genital tract; these organisms may have been *Mycoplasma hominis*. Antiserum inhibited growth in an agar system; in contrast to studies with *M. pneumoniae* (see above), heat inactivation did not seem to inhibit the effect of antiserum. They concluded that the inhibition was independent of complement. Lin and Kass[211] distinguished five different inhibitory effects of antibody on *M. hominis*. Only one of the five, direct killing, required complement. More recently, it has been suggested[212] that C3a, perhaps released from macrophages stimulated by mycoplasma infection, may be toxic to *M. hominis* as well as other mycoplasmas and protoplasts.

C. Other Mycoplasmas

In studies with *M. gallisepticum*, Barker and Patt[213] demonstrated progressive inactivation of the organisms over a 2-hr period in the presence of rabbit antiserum (heat-inactivated) and nonimmune guinea pig serum (1:40 dilution). Neither serum was by itself inhibitory (though more concentrated guinea pig serum [1:10 dilution] was slowly inhibitory over 24 hr). The contribution of the guinea pig serum appeared to be complement, as evidenced by abrogation by heat, absorption with immune complexes, or EDTA.

Killing of *M. canis* by mouse serum also required an intact complement system.[214] The lytic nature of the reaction was suggested by the failure of C5-deficient mouse serum to kill.

D. *Ureaplasma urealyticum*

Also known as "T-strain mycoplasmas," these organisms may be classified as a distinct genus of the family Mycoplasmataceae.[196] They are genital tract mycoplasmas implicated as a cause of urethritis. Lin and Kass[215] found that antibody-induced lysis of these organisms was dependent on or at least greatly enhanced by complement.

E. *Acholeplasma laidlawii*

These organisms are pathogens of animals other than man. Dörner and colleagues[216] found that complement-mediated killing was mediated primarily by IgG antibodies directed against membrane proteins and perhaps phospholipids. The reaction led to release of a cytoplasmic enzyme, hexokinase, and to morphological changes, including rounding, the appearance of membrane "holes", and formation of "ghosts".[217] These changes, typical of complement-mediated lysis, were not associated with disintegration of the membranes. Dahl et al.[218] were able to manipulate changes in serum sensitivity by altering fatty acid content of the cell membrane. They could not relate these changes to simple changes in membrane fluidity.

Original studies indicated that complement alone (nonimmune guinea pig serum) could not produce these membrane lesions.[217] Some mycoplasmacidal activity was subsequently demonstrated in fresh normal guinea pig serum.[219] This activity reflected presence of IgG antibody (perhaps cross-reactive) against mycoplasmal membrane lipids.

F. Comments

Complement can be demonstrated in vivo to participate in opsonization and killing of most mycoplasmas studied. The physiological importance is unknown. But, Sussman et al.[220] reported the intriguing case of a patient with C2 deficiency, recurrent arthritis and arthralgias, and macular and petechial skin rashes. The patient had mycoplasma-like particles observed in serum, but the organisms could not be cultured. These data do not permit any conclusions, even though the purpura was said to have responded to tetracycline therapy.[221] It would be of interest to look further among complement-deficient patients or animals for mycoplasmal infections and vice versa.

XIII. ENDOTOXIN (BACTERIAL LIPOPOLYSACCHARIDE)

The common denominator of all Gram-negative organisms is the outer cell wall of lipopolysaccharides, also called endotoxins for their toxic properties. Experimental injection of endotoxin produces fever, hypotension, endothelial damage, altered blood coagulability, and in extreme cases, circulatory collapse, disseminated intravascular coagulation, and death. The same syndrome is seen clinically in association with Gram-negative bacteremia and sepsis. Most of the symptoms associated with Gram-negative sepsis may be due to concomitant endotoxemia.

Endotoxins (lipopolysaccharides) activate complement quite efficiently and such activation results in lesions on the endotoxin surface in all ways similar to complement-induced erythrocyte surface "holes".[222-224] Contradictory data as to the pathway of complement activated and antibody requirements thereof[225-227] have seemingly been resolved by evidence that both complement pathways can be activated by endotoxin.[223,228-234] The moiety with greatest complement reactivity may be the lipid A portion of the endotoxin molecule,[235-238] although some have thought that lipid A and C1 may interact without further activation of the complement cascade.[234] Morrison and Kline[239] have suggested that lipid A consumes early complement components in an antibody-independent fashion, whereas the polysaccharide portion activates the alternative pathway.

Spink and colleagues[240] suggested in 1964 that complement might be a major mediator of the clinical syndrome of endotoxemia. Dogs developed parallel decrease in complement and increase in histamine following intravenous injection of endotoxin. The investigators suggested that the reaction between endotoxin and natural serum antibodies resulted in complement activation and subsequent histamine release, the histamine mediating shock. Other studies in dogs have confirmed a relationship between the onset of endotoxin shock and hypocomplementemia.[241,242]

Supporting the notion that at least some of the toxic properties of lipopolysaccharide are consequent to complement activation are the demonstrations that (1) irradiation of endotoxin diminished the hypotensive effect in dogs, lethality in mice, and the complement-fixing ability of the molecule;[234] (2) cobra venom factor pretreatment of dogs abolished the early hypotensive response to endotoxin (though later-developing hypotension and mortality were not affected);[243] and (3) cats developed hypotension and complement activation following lipopolysaccharide injection, the former being abrogated by prior cobra venom factor administration.[244] Although Gilbert and Braude[245] measured hypocomplementemia in rabbits after endotoxin injection, Ulevitch et al.[246] noted little change and found that hypotension was not affected by prior treatment with cobra venom factor.

In humans with Gram-negative bacteremia, McCabe[247] showed depressed C3 levels to be associated with shock and death; patients with uncomplicated bacteremia had no definite fall in complement. He could find no relationship to underlying disease, specific organisms, levels of endotoxemia (by limulus lysate tests), or antibody titers. In a prospective study of Gram-negative infections, Füst[248] found a wide variability in complement determinations, but low values tended to correlate with shock. In a few cases, blood was fortuitously obtained 1 or 2 hr prior to onset of shock; in these samples, complement levels were already beginning to decline. Fearon and colleagues[249] found patients with Gram-negative bacteremia to have low levels of properdin, C3, and C1. In each case, values were lower in hypotensive patients than in those who maintained normal blood pressures. Factor B levels were not depressed but tended to be lower in hypotensive patients. Bessa and colleagues[250] found that the increased pulmonary vascular resistance associated with endotoxin injection was due to a serum factor; but, in a model of the isolated perfused canine lung, the factor appeared not to be complement. Endotoxin-induced leukopenia in rabbits or other animals could not be related to complement levels and was not affected by complement depletion.[232,246]

Brown and Lachmann[251] studied the behavior of chromium-labeled rabbit platelets following intravenously administered endotoxin. The majority of platelets were destroyed and all the rabbits eventually died. In contrast, animals depleted of C3–9 survived endotoxin injection and their platelets survived normally. C6-deficient rabbits had the same early thrombocytopenia as did normals, but the platelets eventually returned to the circulation and animals survived. The investigators suggested, therefore, that the lethality of endotoxemia was related somehow to interaction of platelets and complement. Kane et al.[232] showed that C4-deficient guinea pigs did not develop the usual thrombocytopenia and accelerated clotting following endotoxin injection. They suggested that endotoxin combined with platelets, and in the presence of an IgM antibody to endotoxin, the classical pathway was activated with resultant platelet damage, thrombocytopenia, and accelerated clotting. They distinguished this reaction from alternative pathway activation by endotoxin, which they suggested may be antibody independent. The extrapolation of data obtained in animals whose platelets have receptors for C3 should be undertaken cautiously.

For similar reasons, studies of the relationships between endotoxin, complement, and disseminated intravascular coagulation performed in animals with C3 receptor-bearing platelets should be evaluated with caution. Müller-Berghaus and Lohmann[252] suggested that the coagulopathy following intravenous endotoxin injection was not mediated by C6 or later acting components, since the coagulation disturbances were seen in C6-deficient rabbits. Ulevitch et al.[246] comfirmed that finding and demonstrated similar results with cobra venom factor-treated rabbits.

In a rat model, Evensen et al.[253] showed no effect of complement depletion on endothelial injury by endotoxin.

Interaction of endotoxin with serum has been shown to lead to production of an anaphylatoxin, which is a serum factor capable of stimulating histamine release from mast cells.[254] Snyderman and colleagues[224] demonstrated production of a chemotactic factor under similar circumstances.

It is, thus, reasonable to conclude that complement activation by endotoxin may occur in vivo and that such activation may, under appropriate circumstances, lead to some of the pathological sequelae of endotoxemia. Of all the effects of endotoxin in man, hypotension and shock seem most suggestively related to complement activation. But proof of the linkage of endotoxemia to complement activation to hypotension must be considered still hypothetical and perhaps no closer to proven than when Spink[240] postulated it in 1964. Evidence that complement may mediate other effects of endotoxemia, such as disseminated intravascular coagulation, thrombocytopenia, or leukopenia, must be considered even more tenuous.

Some investigators have suggested that, rather than mediating the lethality of endotoxin, complement may provide important protection. Guinea pigs depleted of complement by injection of cobra venom factor and antiserum to C3 had a much shortened survival time following injection of endotoxin.[255] Animals with normal levels of C3 died in approximately 33 hr following injection of a lethal dose of endotoxin; complement-depleted animals died in approximately 11 hr. Johnson and Ward[256] demonstrated that in vitro detoxification of endotoxin by serum plasma decreased mouse lethality, chick embryo lethality, and pyrogenicity in rabbits. They compared the effects of injection of endotoxin into normal rabbits or C6-deficient rabbits.[257] There were no deaths among the normal rabbits, at the dose used in this study; 11 of 12 homozygous C6-deficient rabbits died. They suggested that C6 or one of the later-acting components was important in the detoxification of endotoxin. May et al.[258] studied endotoxin lethality in guinea pigs. C4-deficient guinea pigs depleted of late-acting components with cobra venom factor died much sooner than C4-deficient or cobra venom factor-treated normal guinea pigs, whose survival was only modestly shorter than normal animals. These observations correlated in vitro with impaired detoxifying capacity of C4-deficient, cobra venom factor-treated normal, and cobra venom factor-treated C4-deficient sera.

Studies have not been performed in humans to evaluate the protective role of endotoxin. But it seems possible to interpret the earlier-noted studies showing a relationship between endotoxic shock and hypocomplementemia two ways: on the one hand, complement activation may precipitate shock; on the other hand, it is at least conceivable that complement depletion by continuing endotoxemia exposes an animal to lethal effects of endotoxin. Finally it is entirely possible that complement is activated by endotoxin but has little to do with the clinical syndrome. In that case, low complement levels might be simply a reflection of a particularly high level of endotoxemia.

The generalized Schwartzmann reaction is a laboratory artifact of unknown clinical relevance produced by two sequential intravenous injections of endotoxin, usually into a rabbit. The first (preparative) injection elicits relatively little reaction. The second injection, 24 hr later, is followed by diffuse hemorrhagic necrosis, disseminated intravascular coagulation, and bilateral renal cortical necrosis. It has been suggested that the coagulopathy is due to interactions of platelets and endotoxin. Leukocytes appear to play an important role, as indicated by amelioration of the pathological changes following experimental leukopenia. Heavy deposition of C3 in the renal glomerular lesions, along with fibrin and IgM, might suggest an important role for complement in this reaction.[259]

Stafford and colleagues[260] found no differences in Schwartzmann reactions between normal and C6-deficient rabbits. Likewise, Bergstein and Michael[261] did not alter the pathological changes in rabbits by cobra venom factor pretreatment. In contrast, Fong and Good[262] noted a marked decrease in complement levels following the second injection of endotoxin in rabbits; cobra venom factor treatment of other rabbits aborted both the coagulopathy and pathological changes of the Schwartzmann reaction.

The localized Schwartzmann reaction follows sequential intradermal and (24 hr later) intravenous injection of endotoxin. Hemorrhage occurs at the local site of intradermal injection. Polak and Turk[263] found that both the hemorrhage and leukocyte accumulation could be suppressed by pretreatment of guinea pigs with antiserum to C3.

REFERENCES

1. **Buchanan, R. E. and Gibbons, N. E., Eds.,** *Bergey's Manual of Determinative Bacteriology,* 8th ed., Williams & Wilkins, Baltimore, 1974.
2. **Smith, H.,** Microbial surfaces in relation to pathogenicity, *Bacteriol. Rev.,* 41, 475, 1977.
3. **Leive, L., Ed.,** *Bacterial Membranes and Walls,* Marcel Dekker, New York, 1973.
4. **Mindich, L.,** Synthesis and assembly of bacterial membranes, in *Bacterial Membranes and Walls,* Leive, L., Ed., Marcel Dekker, New York, 1973, 1.
5. **Ghuysen, J.-M. and Shockman, G. D.,** Biosynthesis of peptidoglycan, in *Bacterial Membranes and Walls,* Leive, L., Ed., Marcel Dekker, New York, 1973, 37.
6. **Schleifer, K. H. and Kandler, O.,** Peptidoglycan types of bacterial cell walls and their taxonomic implications, *Bacteriol. Rev.,* 36, 407, 1972.
7. **Archibald, A. R., Baddiley, J., and Blumson, N. L.,** The teichoic acids, *Adv. Enzymol.,* 30, 223, 1968.
8. **Nikaido, H.,** Biosynthesis and assembly of lipopolysaccharide and the outer membrane layer of Gram-negative cell wall, in *Bacterial Membranes and Walls,* Leive, L., Ed., Marcel Dekker, New York, 1973, 131.
9. **Roantree, R. J.,** Salmonella O antigens and virulence, *Annu. Rev. Microbiol.,* 21, 443, 1967.
10. **Orskov, I., Orskov, F., Jann, B., and Jann, K.,** Serology, chemistry, and genetics of O and K antigens of *Escherichia coli, Bacteriol. Rev.,* 41, 667, 1977.
11. **Freeman, B. A.,** The physical and chemical structure of bacteria, in *Burrows Textbook of Microbiology,* 21st ed., W. B. Saunders, Philadelphia, 1979, 13.
12. **MacLeod, C. M. and Krauss, M. R.,** Relation of virulence of pneumococcal strains for mice to the quantitity of capsular polysaccharide formed *in vitro, J. Exp. Med.,* 92, 1, 1950.
13. **Austrian, R. and Gold, J.,** Pneumococcal bacteremia with especial reference to bacteremic pneumococcal pneumonia, *Ann. Intern. Med.,* 60, 759, 1964.
14. **Tramont, E. C.,** *Treponema pallidum* (syphilis), in *Principles and Practice of Infectious Diseases,* Mandell, G. L., Douglas, R. G., Jr., and Bennett, J. E., Eds., John Wiley & Sons, New York, 1979, 1820.
15. **Hederstedt, B.,** Studies on the *Treponema pallidum* immobilizing activity in normal human serum. 2. Serum factors participating in the normal serum immobilization reaction, *Acta Pathol. Microbiol. Scand.,* 84 (c), 135, 1976.
16. **Falls, W. F., Jr., Ford, K. L., Ashworth, C. T., and Carter, N. W.,** The nephrotic syndrome in secondary syphilis. Report of a case with renal biopsy findings, *Ann. Intern. Med.,* 63, 1047, 1965.
17. **Braunstein, G. D., Lewis, E. J., Galvanek, E. G., Hamilton, A., and Bell, W. R.,** The nephrotic syndrome associated with secondary syphilis. An immune deposit disease, *Am. J. Med.,* 48, 643, 1970.
18. **Kaplan, B. S., Wiglesworth, F. W., Marks, M. I., and Drummond, K. N.,** The glomerulopathy of congenital syphilis — an immune deposit disease, *J. Pediatr.,* 81, 1154, 1972.
19. **Bhorade, M. S., Carag, H. B., Lee, H. J., Potter, E. V., and Dunea, G.,** Nephropathy of secondary syphilis. A clinical and pathological spectrum, *JAMA,* 216, 1159, 1971.
20. **Yuceoglu, A. M., Sagel, I., Tresser, G., Wasserman, E., and Lange, K.,** The glomerulopathy of congenital syphilis. A curable immune-deposit disease, *JAMA,* 229, 1085, 1974.
21. **Gamble, C. N. and Reardan, J. B.,** Immunopathogenesis of syphilitic glomerulonephritis. Elution of antitreponemal antibody from glomerular immune-complex deposits, *N. Engl. J. Med.,* 292, 449, 1975.
22. **O'Regan, S., Fong, J. S. C., de Chadarévian, J.-P., Rishikof, J. R., and Drummond, K. M.,** Treponemal antigens in congenital and acquired syphilitic nephritis. Demonstration by immunofluorescence studies, *Ann. Intern. Med.,* 85, 325, 1976.

23. **Fulford, K. W. M., Johnson, N., Loveday, C., Storey, J., and Tedder, R. S.,** Changes in intravascular complement and anti-treponemal antibody titers preceeding the Jarisch-Herxheimer reaction in secondary syphilis, *Clin. Exp. Immunol.,* 24, 483, 1976.
24. **Gelfand, J. A., Elin, R. J., Berry, F. W., Jr., and Frank, M. M.,** Endotoxemia associated with the Jarisch-Herxheimer reaction, *N. Engl. J. Med.,* 295, 211, 1976.
25. **Forsgren, A. and Quie, P. G.,** Influence of the alternate complement pathway on opsonization of several bacterial species, *Infect. Immun.,* 10, 402, 1974.
26. **Bjornson, A. B. and Michael, J. G.,** Factors in human serum promoting phagocytosis of *Pseudomonas aeruginosa*. I. Interaction of opsonins with the bacterium, *J. Infect. Dis.,* 130, S119, 1974.
27. **Gross, G. N., Rehm, S. R., and Pierce, A. K.,** The effect of complement depletion on lung clearance of bacteria, *J. Clin. Invest.,* 62, 373, 1978.
28. **Hann, S. and Holsclaw, D. S.,** Interactions of *Pseudomonas aeruginosa* with immunoglobulins and complement in sputum, *Infect. Immun.,* 14, 114, 1976.
29. **Archer, G. and Fekety, F. R.,** Experimental endocarditis due to *Pseudomonas aeruginosa*. I. Description of a model, *J. Infect. Dis.,* 134, 1, 1976.
30. **Young, L. S. and Armstrong, D.,** Human immunity to *Pseudomonas aeruginosa*. I. In-vitro interaction of bacteria, polymorphonuclear leukocytes, and serum factors, *J. Infect. Dis.,* 126, 257, 1972.
31. **Carlisle, H. N. and Saslaw, S.,** Studies with tularemia vaccines in volunteers. VI. Assessment of role of properdin in resistance, *Proc. Soc. Exp. Biol. Med.,* 110, 603, 1962.
32. **Roantree, R. J. and Rantz, L. A.,** A study of the relationship of the normal bactericidal activity of human serum to bacterial infection, *J. Clin. Invest.,* 39, 72, 1960.
33. **Roantree, R. J. and Pappas, N. C.,** The survival of strains of enteric bacilli in the blood stream as related to their sensitivity to the bactericidal effect of serum, *J. Clin. Invest.,* 39, 82, 1960.
34. **Rowley, D.,** The virulence of strains of *Bacterium coli* for mice, *Br. J. Exp. Pathol.,* 35, 528, 1954.
35. **Olling, S., Hanson, L. A., Holmgren, J., Jodal, U., Lincoln, K., and Lindberg, U.,** The bactericidal effect of normal human serum on *E. coli* strains from normals and from patients with urinary tract infections, *Infection,* 1, 24, 1973.
36. **Miller, T. E., Phillips, S., and Simpson, I. J.,** Complement-mediated immune mechanisms in renal infection. II. Effect of decomplementation, *Clin. Exp. Immunol.,* 33, 115, 1978.
37. **Vosti, K. L. and Randall, E.,** Sensitivity of serologically classified strains of *Escherichia coli* of human origin to the serum bactericidal system, *Am. J. Med. Sci.,* 259, 114, 1970.
38. **Durack, D. T. and Beeson, P. B.,** Protective role of complement in experimental *Escherichia coli* endocarditis, *Infect. Immun.,* 16, 213, 1977.
39. **Howard, C. J. and Glynn, A. A.,** The virulence for mice of strains of *Escherichia coli* related to the effects of K antigens on their resistance to phagocytosis and killing by complement, *Immunology,* 20, 767, 1971.
40. **Björksten, B. and Kaijser, B.,** Interaction of human serum and neutrophils with *Escherichia coli* strains: differences between strains isolated from urine of patients with pyelonephritis or asymptomatic bacteriuria, *Infect. Immun.,* 22, 308, 1978.
41. **van Dijk, W. C., Verbrugh, H. A., van der Tol, M. E., Peters, R., and Verhoef, J.,** Role of *Escherichia coli* K capsular antigens during complement activation, C3 fixation, and opsonization, *Infect. Immun.,* 25, 603, 1979.
42. **Björksten, B., Bortolussi, R., Gothefors, L., and Quie, P. G.,** Interaction of *E. coli* strains with human serum: lack of relationship to K1 antigen, *J. Pediatr.,* 89, 892, 1976.
43. **Bortolussi, R., Ferrieri, P., Björksten, B., and Quie, P. G.,** Capsular K1 polysaccharide of *Escherichia coli*: relationship to virulence in newborn rats and resistance to phagocytosis, *Infect. Immun.,* 25, 293, 1979.
44. **Stevens, P., Huang, S. N.-Y., Welch, W. D., and Young, L. S.,** Restricted complement activation by *Escherichia coli* with the K-1 capsular serotype: a possible role in pathogenicity, *J. Immunol.,* 121, 2174, 1978.
45. **Forsgren, A. and Quie, P. G.,** Opsonic activity in human serum chelated with ethylene glycoltetraacetic acid, *Immunology,* 26, 1251, 1974.
46. **Glovsky, M. M., Alenty, A., and Ghekiere, L.,** Requirement of human C2 and C3 for opsonization and bactericidal activity with human neutrophils, in *Clinical Aspects of the Complement System,* Opferkuch, W., Rother, K., and Schultz, D. R., Eds., Georg Thieme Verlag, Stuttgart, 1978, 152.
47. **Akiyama, Y. and Inoue, K.,** Isolation and properties and complement-resistant strains of *Escherichia coli* K-12, *Infect. Immun.,* 18, 446, 1977.
48. **Guckian, J. C., Christensen, W. D., and Fine, D. P.,** Evidence for quantitative variability of bacterial opsonic requirements, *Infect. Immun.,* 19, 822, 1978.

49. **Leist-Welsh, P. and Bjornson, A. B.**, Requirements for immunoglobulin and the classical and alternative pathways for phagocytosis and intracellular killing of multiple strains of Gram-negative aerobic bacilli, *Infect. Immun.*, 26, 99, 1979.
50. **Gilbert, D. N., Barnett, J. A., and Sanford, J. P.**, *Escherichia coli* bacteremia in the squirrel monkey: demonstration of a complement-dependent neutrophil response, *J. Infect. Dis.*, 128, 251S, 1973.
51. **Gilbert, D. N., Barnett, J. A., and Sanford, J. P.**, *Escherichia coli* bacteremia in the squirrel monkey. I. Effect of cobra venom factor treatment, *J. Clin. Invest.*, 52, 406, 1973.
52. **Rowley, D.**, Sensitivity of rough Gram-negative bacteria to the bactericidal action of serum, *J. Bacteriol*, 95, 1647, 1968.
53. **Reynolds, B. L. and Rowley, D.**, Sensitization of complement resistant bacterial strains, *Nature (London)*, 221, 1259, 1969.
54. **Reynolds, B. L. and Pruul, H.**, Sensitization of complement-resistant smooth Gram-negative bacterial strains, *Infect. Immun.*, 3, 365, 1971.
55. **Reynolds, B. L., Rother, U. A., and Rother, K. O.**, Interaction of complement components with a serum-resistant strain of *Salmonella typhimurium, Infect. Immun.*, 11, 944, 1975.
56. **Bjornson, A. B. and Bjornson, H. S.**, Activation of complement by opportunist pathogens and chemotypes of *Salmonella minnesota, Infect. Immun.*, 16, 748, 1977.
57. **Rother, K. and Rother, U.**, Studies on complement defective rabbits. IV. Blood clearance of intravenously injected *S. typhi* by the reticuloendothelial system, *Proc. Soc. Exp. Biol. Med.*, 119, 1055, 1965.
58. **Schubart, A. F., Hornick, R. B., Ewald, R. W., Schroeder, W. C., Myerburg, R. J., Goodman, J. S., and Woodward, T. E.**, Changes of serum complement and properdin levels in experimental typhoid fever. A longitudinal study of the disease including the incubation period and the phase of recovery following treatment with chloramphenicol, *J. Immunol.*, 93, 387, 1964.
59. **Reed, W. P. and Albright, E. L.**, Serum factors responsible for killing of *Shigella, Immunology*, 26, 205, 1974.
60. **Reed, W. P.**, Serum factors capable of opsonizing *Shigella* for phagocytosis by polymorphonuclear neutrophils, *Immunology*, 28, 1051, 1975.
61. **Wood, W. B., Jr., Smith, M. R., Perry, W. D., and Berry, J. W.**, Studies on the cellular immunology of acute bacteriemia. I. Intravascular leucocytic reaction and surface phagocytosis, *J. Exp. Med.*, 94, 521, 1951.
62. **Simberkoff, M. S., Ricupero, I., and Rahal, J. J., Jr.**, Host resistance to *Serratia marcescens* infection: serum bactericidal activity and phagocytosis by normal blood leukocytes, *J. Lab. Clin. Med.*, 87, 206, 1976.
63. **Traub, W. H. and Kleber, I.**, Selective activation of classical and alternative pathways of human complement by "promptly serum-sensitive" and "delayed serum-sensitive" strains of *Serratia marcescens, Infect. Immun.*, 13, 1343, 1976.
64. **Fine, D. P., Marney, S. R., Jr., Colley, D. G., and Des Prez, R. M.**, *Hemophilus influenzae* decomplementation pattern in chelated and nonchelated serum, in *Hemophilus influenzae*, Sell, S. H. W. and Karzon, D. T., Eds., Vanderbilt University Press, Nashville, 1973, 113.
65. **Quinn, P. H., Crosson, F. J., Jr., Winkelstein, J. A., and Moxon, E. R.**, Activation of the alternative complement pathway by *Hemophilus influenzae* type b, *Infect. Immun.*, 16, 400, 1977.
66. **Crosson, F. J., Jr., Winkelstein, J. A., and Moxon, E. R.**, Participation of complement in the nonimmune host defense against experimental *Haemophilus influenzae* type b septicemia and meningitis, *Infect. Immun.*, 14, 882, 1976.
67. **Weller, P. F., Smith, A. L., Smith, D. H., and Anderson, P.**, Role of immunity in the clearance of bacteremia due to *Haemophilus influenzae, J. Infect. Dis.*, 138, 427, 1978.
68. **Newman, S. L., Waldo, B., and Johnston, R. B., Jr.**, Separation of serum bactericidal and opsonizing activities for *Haemophilus influenzae* type b, *Infect. Immun.*, 8, 488, 1973.
69. **Johnston, R. B., Jr., Anderson, P., and Newman, S. L.**, Opsonization and phagocytosis of *Hemophilus influenzae*, type b, in *Hemophilus influenzae*, Sell, S. H. W. and Karzon, D. T., Eds., Vanderbilt University Press, Nashville, 1973, 99.
70. **Casciato, D. A., Rosenblatt, J. E., Goldberg, L. S., and Bluestone, R.**, In vitro interaction of *Bacteroides fragilis* with polymorphonuclear leukocytes and serum factors, *Infect. Immun.*, 11, 337, 1975.
71. **Casciato, D. A., Rosenblatt, J. E., Bluestone, R., Goldberg, L. S., and Finegold, S. M.**, Susceptibility of isolates of *Bacteroides* to the bactericidal activity of normal human serum, *J. Infect. Dis.*, 140, 109, 1979.

72. **Bjornson, A. B., Altemeier, W. A., and Bjornson, H. S.**, Comparison of the *in vitro* bactericidal activity of human serum and leukocytes against *Bacteroides fragilis* and *Fusobacterium mortiferum* in aerobic and anaerobic environments, *Infect. Immun.*, 14, 843, 1976.
73. **Bjornson, A. B. and Bjornson, H. S.**, Participation of immunoglobulin and the alternative complement pathway in opsonization of *Bacteroides fragilis* and *Bacteroides thetaiotaomicron, J. Infect. Dis.*, 138, 351. 1978.
74. **Sveen, K.**, The importance of C5 and the role of the alternative complement pathway in leukocyte chemotaxis induced *in vivo* and *in vitro* by *Bacteroides fragilis* lipopolysaccharide, *Acta Pathol. Microbiol. Scand. Sec. B*, 86, 93, 1978.
75. **Nygren, H., Dahlen, G., and Nilsson, L.-A.**, Human complement activation by lipopolysaccharides from *Bacteroides oralis, Fusobacterium nucleatum,* and *Veillonella parvula, Infect. Immun.*, 26, 391, 1979.
76. **Hawley, C. E. and Falkler, W. A., Jr.**, Anticomplementary activity of *Fusobacterium polymorphum* in normal and C4-deficient sources of guinea pig complement, *Infect. Immun.*, 18, 124, 1977.
77. **Brooks, G. F., Israel, K. S., and Petersen, B. H.**, Bactericidal and opsonic activity against *Neisseria gonorrhoeae* in sera from patients with disseminated gonococcal infection, *J. Infect. Dis.*, 134, 450, 1976.
78. **Schoolnik, G. K., Buchanan, T. M., and Holmes, K. K.**, Gonococci causing disseminated gonococcal infection are resistant to the bactericidal action of normal human sera, *J. Clin. Invest.*, 58, 1163, 1976.
79. **Eisenstein, B. I., Lee, T. J., and Sparling, P. F.**, Penicillin sensitivity and serum resistance are independent attributes of strains of *Neissera gonorrhoeae* causing disseminated gonococcal infection, *Infect. Immun.*, 15, 834, 1977.
80. **Brooks, G. F. and Ingwer, I.**, Studies on the relationships between serum bactericidal activity and uncomplicated genital infections due to *Neisseria gonorrhoeae, J. Infect. Dis.*, 138, 333, 1978.
81. **Tramont, E. C., Sadoff, J. C., and Wilson, C.**, Varability of the lytic susceptibility of *Neisseria gonorrhoeae* to human sera, *J. Immunol.*, 118, 1843, 1977.
82. **Ingwer, I., Petersen, B. H., and Brooks, G.**, Serum bactericidal action and activation of the classic and alternate complement pathways by *Neisseria gonorrhoeae, J. Lab. Clin. Med.*, 92, 211, 1978.
83. **Arko, R. J., Wong, K. H., Steurer, F. J., and Schalla, W. O.**, Complement-enhanced immunity to infection with *Neisseria gonorrhoeae* in mice, *J. Infect. Dis.*, 139, 569, 1979.
84. **Schiller, N. L., Friedman, G. L., and Roberts, R. B.**, The role of natural IgG and complement in the phagocytosis of type 4 *Neiserria gonorrhoeae* by human polymorphonuclear leukocytes, *J. Infect. Dis.*, 140, 698, 1979.
85. **Thomas, L. and Dingle, J. H.**, Investigation of meningococcal infection. I. Bacteriological aspects, *J. Clin. Invest.*, 22, 353, 1943.
86. **Johnson, U. and Laurell, A.-B.**, Complement activation in meningococcal septicemia, *Acta Pathol. Microbiol. Scand. Sec. C*, 83, 285, 1975.
87. **Hoffman, T. A. and Edwards, E. A.**, Group-specific polysaccharide antigen and humoral antibody response in disease due to *Neisseria meningitidis, J. Infect. Dis.*, 126, 636, 1972.
88. **Greenwood, B. M., Onyewotu, I. I., and Whittle, H. C.**, Complement and meningococcal infection, *Br. Med. J.*, 1, 797, 1976.
89. **Black, J. A., Challacombe, D. N., and Ockenden, B. G.**, Nephrotic syndrome associated with bacteremia after shunt operations for hydrocephalus, *Lancet*, 2, 921, 1965.
90. **Stickler, G. B., Shin, M. H., Burke, E. C., Holley, K. E., Miller, R. H., and Segar, W. E.**, Diffuse glomerulonephritis associated with infected ventriculoatrial shunt, *N. Engl. J. Med.*, 279, 1077, 1968.
91. **Lam, C. N., McNeish, A. S., and Gibson, A. A. M.**, Nephrotic syndrome associated with complement deficiency and *Staphylococcus albus* bacteraemia, *Scott. Med. J.*, 14, 86, 1969.
92. **Rames, L., Wise, B., Goodman, J. R., and Piel, C. F.**, Renal disease with *Staphylococcus albus* bacteremia. A complication in ventriculoatrial shunts, *JAMA*, 212, 1671, 1970.
93. **Kaufman, D. B. and McIntosh, R.**, The pathogenesis of the renal lesion in a patient with streptococcal disease, infected ventriculoatrial shunt, cryoglobulinemia and nephritis, *Am. J. Med.*, 50, 262, 1971.
94. **Dobrin, R. S., Day, N. K., Quie, P. G., Moore, H. L., Vernier, R. L., Michael, A. F., and Fish, A. J.**, The role of complement, immunoglobulin and bacterial antigen in coagulase-negative staphylococcal shunt nephritis, *Am. J. Med.*, 59, 660, 1975.
95. **Strife, C. F., McDonald, B. M., Ruley, E. J., McAdams, A. J., and West, C. D.**, Shunt nephritis: the nature of the serum cryoglobulins and their relation to the complement profile, *J. Pediatr.*, 88, 403, 1976.

96. **Schoenbaum, S. C., Gardner, P., and Shillito, J.,** Infections of cerebrospinal fluid shunts: epidemiology, clinical manifestations, and therapy, *J. Infect. Dis.,* 131, 543, 1975.
97. **Cohn, Z. A. and Morse, S. I.,** Interactions between rabbit polymorphonuclear leukocytes and staphylococci, *J. Exp. Med.,* 110, 419, 1959.
98. **Verhoef, J., Peterson, P. K., Kim, Y., Sabath, L. D., and Quie, P. G.,** Opsonic requirements for staphylococcal phagocytosis: heterogeneity among strains, *Immunology,* 33, 191, 1977.
99. **Wheat, L. J., Humphreys, D. W., and White, A.,** Opsonization of staphylococci by normal human sera: the role of antibody and heat-labile factors, *J. Lab. Clin. Med.,* 83, 73, 1974.
100. **Guckian, J. C., Christensen, W. D., and Fine, D. P.,** Trypan blue inhibits complement-mediated phagocytosis by human polymorphonuclear leukocytes, *J. Immunol.,* 120, 1580, 1978.
101. **Verbrugh, H. A., van Dijk, W. C., Peters, R., van der Tol, M. E., Peterson, P. K., and Verhoef, J.,** *Staphylococcus aureus* opsonization mediated via the classical and alternative complement pathways. A kinetic study using MgEGTA chelated serum and human sera deficient in IgG and complement factors C1s and C2, *Immunology,* 36, 391, 1979.
102. **Murphey, S. A., Root, R. K., and Schreiber, A. D.,** The role of antibody and complement in phagocytosis by rabbit alveolar macrophages, *J. Infect. Dis.,* 140, 896, 1979.
103. **Peterson, P. K., Wilkinson, B. J., Kim, Y., Schmeling, D., and Quie, P. G.,** Influence of encapsulation on staphylococcal opsonization and phagocytosis by human polymorphonuclear leukocytes, *Infect. Immun.,* 19, 943, 1978.
104. **Peterson, P. K., Kim, Y., Wilkinson, B. J., Schmeling, D., Michael, A. F., and Quie, P. G.,** Dichotomy between opsonization and serum complement activation by encapsulated staphylococci, *Infect. Immun.,* 20, 770, 1978.
105. **Koenig, M. G., Melly, M. A., Goodman, J. S., and Rogers, D. E.,** Serum factors and the reticuloendothelial uptake of *Staphylococcus aureus*. I. The role of whole serum, *J. Immunol.,* 100, 516, 1968.
106. **Fine, D., Guckian, J., Harper, B., Christensen, G., and Daniels, J.,** Opsonisation of pneumococci, in *Pathogenic Streptococci,* Parker, M. T., Ed., Reedbooks Ltd., Chertsey, Surrey, 1979, 193.
107. **Wilkinson, B. J., Peterson, P. K., and Quie, P. G.,** Cryptic peptidoglycan and the antiphagocytic effect of the *Staphylococcus aureus* capsule: model for the antiphagocytic effect of bacterial cell surface polymers, *Infect. Immun.,* 23, 502, 1979.
108. **Wilkinson, B. J., Sisson, S. P., Kim, Y., and Peterson, P. K.,** Localization of the third component of complement on the cell wall of encapsulated *Staphylococcus aureus* M: implications for the mechanism of resistance to phagocytosis, *Infect. Immun.,* 26, 1159, 1979.
109. **Verbrugh, H. A., van Dijk, W. C., van Erne, M. E., Peters, R., Peterson, P. K., and Verhoef, J.,** Quantitation of the third component of human complement attached to the surface of opsonized bacteria: opsonin-deficient sera and phagocytosis-resistant strains, *Infect. Immun.,* 26, 808, 1979.
110. **Saulsbury, F. T. and Winkelstein, J. A.,** Activation of the alternative complement pathway by L-phase variants of Gram-positive bacteria, *Infect. Immun.,* 23, 711, 1979.
111. **Shayegani, M., Hisatsune, K., and Mudd, S.,** Cell wall component which affects the ability of serum to promote phagocytosis and killing of *Staphylococcus aureus, Infect. Immun.,* 2, 750, 1970.
112. **Peterson, P. K., Wilkinson, B. J., Kim, Y., Schmeling, D., Douglas, S. D., Quie, P. G., and Verhoef, J.,** The key role of peptidoglycan in the opsonization of *Staphylococcus aureus, J. Clin. Invest.,* 61, 597, 1978.
113. **Schmeling, D. J., Peterson, P. K., Hammerschmidt, D. E., Kim, Y., Verhoef, J., Wilkinson, B. J., and Quie, P. G.,** Chemotaxigenesis by cell surface components of *Staphylococcus aureus, Infect. Immun.,* 26, 57, 1979.
114. **Sjöquist, J. and Stalenheim, G.,** Protein A from *Staphylococcus aureus*. IX. Complement-fixing activity of protein A-IgG complexes, *J. Immunol.,* 103, 467, 1969.
115. **Stalenheim, G., Götze, O., Cooper, N. R., Sjöquist, J., and Müller-Eberhard, H. J.,** Consumption of human complement components by complexes of IgG with protein A of *Staphylococcus aureus, Immunochemistry,* 10, 501, 1973.
116. **Forsgren, A. and Quie, P. G.,** Effects of staphyloccal protein A on heat labile opsonins, *J. Immunol.,* 112, 1177, 1974.
117. **Peterson, P. K., Verhoef, J., Sabath, L. D., and Quie, P. G.,** Effect of protein A on staphylococcal opsonization, *Infect. Immun.,* 15, 760, 1977.
118. **Easmon, C. S. F. and Glynn, A. A.,** Comparision of subcutaneous and intraperitoneal staphylococcal infections in normal and complement-deficient mice, *Infect. Immun.,* 13, 399, 1976.
119. **Medhurst, F. A., Hill, M. J., and Glynn, A. A.,** The effect of antilymphocyte serum on subcutaneous staphylococcal infections in normal, immune and complement-deficient mice, *J. Med. Microbiol.,* 2, 147, 1969.

120. **Duma, R. J., Weinberg, A. N., Medrek, T. F., and Kunz, L. J.,** Streptococcal infections. A bacteriologic and clinical study of streptococcal bacteremia, *Medicine*, 48, 87, 1969.
121. **Keefer, C. S. and Spink, W. W.,** Studies of hemolytic streptococcal infection. I. Factors influencing the outcome of erysipelas, *J. Clin. Invest.*, 15, 17, 1936.
122. **Spink, W. W. and Keefer, C. S.,** Studies of hemolytic streptococcal infection. II. The serological reactions of the blood during erysipelas, *J. Clin. Invest.*, 15, 21, 1936.
123. **Greenblatt, J., Boackle, R. J., and Schwab, J. H.,** Activation of the alternate complement pathway by peptidoglycan from streptococcal cell wall, *Infect. Immun.*, 19, 296, 1978.
124. **Stollerman, G. H., Alberti, H., and Plemmons, J. A.,** Opsonization of group A streptococci by complement deficient blood from a patient with hereditary angioneurotic edema, *J. Immunol.*, 99, 92, 1967.
125. **Sakai, S., Ryovama, K., Koshimura, S., and Migita, S.,** Studies on the properties of a streptococcal preparation OK-432 (NSC-B116209) as an immunopotentiator. I. Activation of serum complement components and peritoneal exudate cells by group A streptococcus, *Jpn. J. Exp. Med.*, 46, 123, 76.
126. **Hemming, V. G., Hall, R. T., Rhodes, P. G., Shigeoka, A. O., and Hill, H. R.,** Assessment of group B streptococcal opsonins in human and rabbit serum by neutrophil chemiluminescence, *J. Clin. Invest.*, 58, 1379, 1976.
127. **Mathews, J. H., Klesius, P. H., and Zimmerman, R. A.,** Opsonin system of the group B streptococcus, *Infect. Immun.*, 10, 1315, 1974.
128. **Resnick, G. D. and Becker, C. G.,** Bypass activation of complement by streptococcal cell walls, *Circulation*, 48(Suppl. IV), 80, 1973.
129. **Inai, S., Nagaki, K., Ebisu, S., Kato, K., Kotani, S., and Misaki, A.,** Activation of the alternative complement pathway by water-insoluble glucans of *Streptococcus mutans:* the relation between their chemical structures and activating potencies, *J. Immunol.*, 117, 1256, 1976.
130. **Silvestri, L. J., Knox, K. W., Wicken, A. J., and Hoffman, E. M.,** Inhibition of complement-mediated lysis of sheep erythrocytes by cell-free preparations from *Streptococcus mutans* BHT, *J. Immunol.*, 122, 54, 1979.
131. **Tauber, J. W., Polley, M. J., and Zabriskie, J. B.,** Nonspecific complement activation by streptococcal structures. II. Properdin-independent initiation of the alternate pathway, *J. Exp. Med.*, 143, 1352, 1976.
132. **Peterson, P. K., Schmeling, D., Cleary, P. P., Wilkinson, B. J., Kim, Y., and Quie, P. G.,** Inhibition of alternative complement pathway opsonization by group A streptococcal M protein, *J. Infect. Dis.*, 139, 575, 1979.
133. **Bisno, A. L.,** Alternate complement pathway activation by group A streptococci: role of M-protein, *Infect. Immun.*, 26, 1172, 1979.
134. **Kronvall, G.,** A surface component in group A, C, and G streptococci with non-immune reactivity for immunoglobulin G, *J. Immunol.*, 111, 1401, 1973.
135. **Myhre, E. B. and Kronvall, G.,** Heterogeneity of nonimmune immunoglobulin Fc reactivity among Gram-positive cocci: description of three major types of receptors for human immunoglobulin G, *Infect. Immun.*, 17, 475, 1977.
136. **Michael, A. F., Jr., Drummond, K. N., Good, R. A., Vernier, R. L.,** Acute poststreptococcal glomerulonephritis: immune deposit disease, *J. Clin. Invest.*, 45, 237, 1966.
137. **Feldman, J. D., Mardiney, M. R., and Shuler, S. E.,** Immunology and morphology of acute poststreptococcal glomerulonephritis, *Lab. Invest.*, 15, 283, 1966.
138. **Fischel, E. E. and Gajdusek, D. C.,** Serum complement in acute glomerulonephritis and other renal diseases, *Am. J. Med.*, 12, 190, 1952.
139. **Levy, R. L. and Hong, R.,** The immune nature of subacute bacterial endocarditis (SBE) nephritis, *Am. J. Med.*, 54, 645, 1973.
140. **Morel-Maroger, L., Sraer, J.-D., Herreman, G., and Godeau, P.,** Kidney in subacute endocarditis. Pathological and immunofluorescence findings, *Arch. Pathol.*, 94, 205, 1972.
141. **Arnold, S. B.. Valone, J. A., Askenase, P. W., Kashgarian, M., and Freedman, L. R.,** Diffuse glomerulonephritis in rabbits with *Streptococcus viridans* endocarditis, *Lab. Invest.*, 32, 681, 1975.
142. **Lund, E.,** Types of pneumococci found in blood, spinal fluid and pleural exudate during a period of 15 years (1954–1969), *Acta Pathol. Microbiol. Scand. Sec. B*, 78, 333, 1970.
143. **Mufson, M. A., Kruss, D. M., Wasil, R. E., and Metzger, W. I.,** Capsular types and outcome of bacteremic pneumococcal disease in the antibiotic era, *Arch. Intern. Med.*, 134, 505, 1974.
144. **Dick, G. F.,** On the development of proteolytic ferments in the blood during pneumonia, *J. Infect. Dis.*, 10, 383, 1912.

145. **Rutstein, D. D. and Walker, W. H.,** Complement activity in pneumonia, *J. Clin. Invest.*, 21, 347, 1942.
146. **Reed, W. P., Davidson, M. S., and Williams, R. C., Jr.,** Complement system in pneumococcal infections, *Infect. Immun.*, 13, 1120, 1976.
147. **Coonrod, J. D. and Rylko-Bauer, B.,** Complement levels in pneumococcal pneumonia, *Infect. Immun.*, 18, 14, 1977.
148. **Guckian, J. C., Christensen, G. D., and Fine, D. P.,** The role of opsonins in recovery from experimental pneumococcal pneumonia, *J. Infect. Dis.*, 142, 175, 1980.
149. **Young, L. H., Maroko, P. R., Carpenter, C. B., and Murphy, J.,** Reduction of the extent of experimental pneumonia by cobra venom factor administration, *Clin. Res.*, 23, 299A, 1975.
150. **Robertson, O. H., Hamburger, M., Jr., and Gregg, L. A.,** On the nature of bacteremia in experimental pneumococcal pneumonia in the dog. I. Relationship of natural pneumococcidal-promoting activity of the serum to blood invasion, *J. Exp. Med.*, 97, 283, 1953.
151. **Gregg, L. A. and Robertson, O. H.,** On the nature of bacteremia in experimental pneumococcal pneumonia in the dog. II. Disappearance of pneumococci from the circulation in relation to the bactericidal action of the blood *in vitro*, *J. Exp. Med.*, 97, 297, 1953.
152. **Ward, H. K. and Enders, J. F.,** An analysis of the opsonic and tropic action of normal and immune serum based on experiments with the pneumococcus, *J. Exp. Med.*, 57, 527, 1933.
153. **Jeter, W. S., McKee, A. P., and Mason, R. J.,** Inhibition of immune phagocytosis of *Diplococcus pneumoniae* by human neutrophiles with antibody against complement, *J. Immunol.*, 86, 386, 1961.
154. **Fine, D. P.,** Pneumococcal type-associated variability in alternate complement pathway activation, *Infect. Immun.*, 12, 772, 1975.
155. **Giebink, G. S., Verhoef, J., Peterson, P. K., and Quie, P. G.,** Opsonic requirements for phagocytosis of *Streptococcus pneumoniae* types VI, XVIII, XXIII, and XXV, *Infect. Immun.*, 18, 291, 1977.
156. **Coonrod, J. D. and Jenkins, S.,** Interaction of pneumococcal antigens with complement in rats, *Infect. Immun.*, 23, 626, 1979.
157. **Smith, M. R. and Wood, W. B., Jr.,** Heat labile opsonins to pneumococcus. I. Participation of complement, *J. Exp. Med.*, 130, 1209, 1969.
158. **Stephens, C. G., Williams, R. C., Jr., and Reed, W. P.,** Classical and alternative complement pathway activation by pneumococci, *Infect. Immun.*, 17, 296, 1977.
159. **Winkelstein, J. A., Bocchini, J. A., Jr., and Schiffman, G.,** The role of the capsular polysaccharide in the activation of the alternative pathway by the pneumococcus, *J. Immunol.*, 116, 367, 1976.
160. **Winkelstein, J. A., Smith, M. R., and Shin, H. S.,** The role of C3 as an opsonin in the early stages of infection, *Proc. Soc. Exp. Biol. Med.*, 149, 397, 1975.
161. **Winkelstein, J. A., Shin, H. S., and Wood, W. B., Jr.,** Heat labile opsonins to pneumococcus. III. The participation of immunoglobulin and of the alternate pathway of C3 activation, *J. Immunol.*, 108, 1681, 1972.
162. **Dhingra, R. K., Williams, R. C., Jr., and Reed, W. P.,** Effects of pneumococcal mucopeptide and capsular polysaccharide on phagocytosis, *Infect. Immun.*, 15, 169, 1977.
163. **Winkelstein, J. A. and Tomasz, A.,** Activation of the alternative pathway by pneumococcal cell walls, *J. Immunol.*, 118, 451, 1977.
164. **Winkelstein, J. A. and Tomasz, A.,** Activation of the alternative pathway by pneumococcal cell wall teichoic acid, *J. Immunol.*, 120, 174, 1978.
165. **Stephens, C. G., Reed, W. P., Kronvall, G., and Williams, R. C., Jr.,** Reactions between certain strains of pneumococci and Fc of IgG, *J. Immunol.*, 112, 1955, 1974.
166. **Myhre, E. B. and Kronvall, G.,** Heterogeneity of nonimmune immunoglobulin Fc reactivity among Gram-positive cocci: descriptions of three major types of receptors for human immunoglobulin G, *Infect. Immun.*, 17, 475, 1977.
167. **Hyman, L. R., Jenis, E. H., Hill, G. S., Zimmerman, S. W., and Burkholder, P. M.,** Alternate C3 pathway activation in pneumococcal glomerulonephritis, *Am. J. Med.*, 58, 810, 1975.
168. **Baker, L. A., Campbell, P. A., and Hollister, J. R.,** Chemotaxigenesis and complement fixation by *Listeria monocytogenes* cell wall fractions, *J. Immunol.*, 119, 1723, 1977.
169. **Dinter, Z., Diderholm, H., and Rockborn, G.,** Complement-dependent hemolysis following hemagglutination by *Erysipelothrix rhusiopathiae*, *Zentralbl. Bakteriol.*, [Orig. A] 236, 533, 1976.
170. **Timoney, J.,** The effect of decomplementation on *Erysipelothrix rhusiopathiae* infection in the mouse, *Immunology*, 19, 561, 1970.
171. **Caren, L. D. and Rosenberg, L. T.,** The role of complement in resistance to endogenous and exogenous infection with a common mouse pathogen, *Corynebacterium kutscheri*, *J. Exp. Med.*, 124, 689, 1966.

172. **Bolton, W. K., Sande, M. E., Normansell, D. E., Sturgill, B. C., and Westervelt, F. B., Jr.,** Ventriculojugular shunt nephritis with *Corynebacterium bovis*. Successful therapy with antibiotics, *Am. J. Med.*, 59, 417, 1975.
173. **Moss, S. W., Gary, N. E., and Eisinger, R. P.,** Nephritis associated with a diphtheroid-infected cerebrospinal fluid shunt, *Am. J. Med.*, 63, 318, 1977.
174. **Biran, H., Moake, J. L., Reed, R. C., Gutterman, J. U., Hersh, E. M., Freireich, E. J., and Mavligit, G. M.,** Complement activation *in vivo* in cancer patients receiving *C. parvum* immunotherapy, *Br. J. Cancer*, 34, 493, 1976.
175. **McBride, W. H., Weir, D. M., Kay, A. B., Pearce, P., and Caldwell, J. R.,** Activation of the classical and alternate pathways of complement by *Corynebacterium parvum*, *Clin. Exp. Immunol.*, 19, 143, 1975.
176. **Webster, G. F., Leyden, J. J., Norman, M. E., and Nilsson, U. R.,** Complement activation in acne vulgaris: *in vitro* studies with *Propionibacterium acnes* and *Propionibacterium granulosum*, *Infect. Immun.*, 22, 523, 1978.
177. **Webster, G. F., Leyden, J. J., and Nilsson, U. R.,** Complement activation in acne vulgaris: consumption of complement by comedones, *Infect. Immun.*, 26, 183, 1979.
178. **Beeler, B. A., Crowder, J. G., Smith, J. W., and White, A.,** *Propionibacterium acnes:* pathogen in central nervous system shunt infection. Report of three cases including immune complex glomerulonephritis, *Am. J. Med.*, 61, 935, 1976.
179. **Rourke, F. J., Fan, S. S., and Wilder, M. S.,** Anticomplementary activity of tuberculin: relationship to platelet aggregation and lytic response, *Infect. Immun.*, 23, 160, 1979.
180. **Wemambu, S. N. C., Turk, J. L., Waters, M. F. R., and Rees, R. J. W.,** Erythema nodosum leprosum: a clinical manifestation of the Arthus phenomenon, *Lancet*, 2, 933, 1969.
181. **de Azevedo, M. P. and de Melo, P. H.,** A comparative study of the complementary activity of serum in the polar forms of leprosy and in the leprosy reaction, *Int. J. Leprosy*, 34, 34, 1966.
182. **Saitz, E. W., Dierks, R. E., and Shepard, C. C.,** Complement and the second component of complement in leprosy, *Int. J. Leprosy*, 36, 400, 1968.
183. **Petchclai, B., Chutanondh, R., Prasongsom, S., Hiranras, S., and Ramasoota, T.,** Complement profile in leprosy, *Am. J. Trop. Med. Hyg.*, 22, 761, 1973.
184. **Srivastava, L. M., Agarwal, D. P., and Goedde, H. W.,** Biochemical, immunological and genetic studies in leprosy. II. Profile of immunoglobulins, complement components and C-reactive protein in sera of leprosy patients and healthy controls, *Tropenmed. Parasitol.*, 26, 212, 1975.
185. **Srivastava, L. M., Agarwal, D. P., Benkmann, H. G., and Goedde, H. W.,** Biochemical, immunological and genetic studies in leprosy. III. Genetic polymorphism of C3 and immunoglobulin profile in leprosy patients, healthy family members and controls, *Tropenmed. Parasitol.*, 26, 426, 1975.
186. **Edwards, J. H.,** A quantitative study on the activation of the alternative pathway of complement by mouldy hay dust and thermophilic actinomycetes, *Clin. Allergy*, 6, 19, 1976.
187. **Edwards, J. H., Baker, J. T., and Davies, B. H.,** Precipitin test negative farmer's lung—activation of the alternative pathway of complement by mouldy hay dusts, *Clin. Allergy*, 4, 379, 1974.
188. **Marx, J. J., Jr. and Flaherty, D. K.,** Activation of the complement sequence by extracts of bacteria and fungi associated with hypersensitivity pneumonitis, *J. Allergy Clin. Immunol.*, 57, 328, 1976.
189. **Burrell, R. and McCullough, M. J.,** Production of thermophilic actinomycete-hay aerosols for use in experimental hypersensitivity pneumonitis, *Appl. Environ. Microbiol.*, 34, 715, 1977.
190. **Hand, W. L., Miller, J. B., Reinarz, J. A., and Sanford, J. P.,** Rocky Mountain spotted fever. A vascular disease, *Arch. Intern. Med.*, 125, 879, 1970.
191. **Graybill, J. R., Hawiger, J., and Des Prez, R. M.,** Complement and coagulation in Rocky Mountain spotted fever, *South. Med. J.*, 66, 410, 1973.
192. **Moe, J. B., Mosher, D. F., Kenyon, R. H., White, J. D., Stookey, J. L., Bagley, L. R., and Fine, D. P.,** Functional and morphologic changes during experimental Rocky Mountain spotted fever in guinea pigs, *Lab. Invest.*, 35, 235, 1976.
193. **de Brito, T., Hoshino-Shimizu, S., Pereira, M. O., and Rigolon, N.,** The pathogenesis of the vascular lesions in experimental rickettsial disease of the guinea pig (Rocky Mountain spotted fever group). A light, immunofluorescent and electron microscopic study, *Virchows Arch. A*, 358, 205, 1973.
194. **Mosher, D. F., Fine, D. P., Moe, J. B., Kenyon, R. H., and Ruch, G. L.,** Studies of the coagulation and complement systems during experimental Rocky Mountain spotted fever in rhesus monkeys, *J. Infect. Dis.*, 135, 985, 1977.
195. **Fine, D., Mosher, D., Yamada, T., Burke, D., and Kenyon, R.,** Coagulation and complement studies in Rocky Mountain spotted fever, *Arch. Intern. Med.*, 138, 735, 1978.

196. **Couch, R. B.,** Mycoplasma diseases. Introduction, in *Principles and Practice of Infectious Diseases,* Mandell, G. L., Douglas, R. G., Jr., and Bennett, J. E., Eds., John Wiley & Sons, New York, 1979, 1481.
197. **Muschel, L. H. and Jackson, J. E.,** The reactivity of serum against protoplasts and spheroplasts, *J. Immunol.,* 97, 46, 1966.
198. **Couch, R. B.,** *Mycoplasma pneumoniae* (primary atypical pneumonia), in *Principles and Practice of Infectious Diseases,* Mandell, G. L., Douglas, R. G., Jr., and Bennett, J. E., John Wiley & Sons, New York, 1979, 1484.
199. **Eaton, M. D., Farnham, A. E., Levinthal, J. D., and Scala, A. R.,** Cytopathic effect of the atypical pneumonia organism in cultures of human tissue, *J. Bacteriol.,* 84, 1330, 1962.
200. **Taylor-Robinson, D., Purcell, R. H., Wong, D. C., and Chanock, R. M.,** A colour test for the measurement of antibody to certain mycoplasma species based upon the inhibition of acid production, *J. Hyg.,* 64, 91, 1966.
201. **Fernald, G. W., Clyde, W. A., Jr., and Denny, F. W.,** Factors influencing growth inhibition of *Mycoplasma pneumoniae* by immune sera, *Proc. Soc. Exp. Biol. Med.,* 126, 161, 1967.
202. **Gale, J. L. and Kenny, G. E.,** Complement dependent killing of *Mycoplasma pneumoniae* by antibody: kinetics of the reaction, *J. Immunol.,* 104, 1175, 1970.
203. **Brunner, H., James, W. D., Horswood, R. L., and Chanock, R. M.,** Measurement of *Mycoplasma pneumoniae* mycoplasmacidal antibody in human serum, *J. Immunol.,* 108, 1491, 1972.
204. **Brunner, H., Razin, S., Kalica, A. R., and Chanock, R. M.,** Lysis and death of *Mycoplasma pneumoniae* by antibody and complement, *J. Immunol.,* 106, 907, 1971.
205. **Brunner, H., Kalica, A. R., James, W. D., Horswood, R. L., and Chanock, R. M.,** Ultrastructural lesions in *Mycoplasma pneumoniae* membranes produced by antibody and complement, *Infect. Immun.,* 7, 259, 1973.
206. **Bredt, W. and Bitter-Suermann, D.,** Interactions between *Mycoplasma pneumoniae* and guinea pig complement, *Infect. Immun.,* 11, 497, 1975.
207. **Bredt, W.,** Phagocytosis by macrophages of *Mycoplasma pneumoniae* after opsonization by complement, *Infect. Immun.,* 12, 694, 1975.
208. **Bredt, W., Wellek, B., Brunner, H., and Loos, M.,** Interactions between *Mycoplasma pneumoniae* and the first components of complement, *Infect. Immun.,* 15, 7, 1977.
209. **Loos, M. and Brunner, H.,** Complement components (C1, C2, C3, C4) in bronchial secretions after intranasal infection of guinea pigs with *Mycoplasma pneumoniae*: dissociation of unspecific and specific defense mechanisms, *Infect. Immun.,* 25, 583, 1979.
210. **Edward, D. G. and Fitzgerald, W. A.,** Inhibition of growth of pleuropneumonia-like organisms by antibody, *J. Pathol. Bacteriol.,* 68, 23, 1954.
211. **Lin, J.-S. and Kass, E. H.,** Complement-dependent and complement-independent interactions between *Mycoplasma hominis* and antibodies *in vitro, J. Med. Microbiol.,* 8, 397, 1975.
212. **Taylor-Robinson, D., Schorlemmer, H. U., Furr, P. M., and Allison, A. C.,** Macrophage secretion and the complement cleavage product C3a in the pathogenesis of infections by mycoplasmas and L-forms of bacteria and in immunity to these organisms, *Clin. Exp. Immunol.,* 33, 486, 1978.
213. **Barker, L. F. and Patt, J. K.,** Role of complement in immune inactivation of *Mycoplasma gallisepticum, J. Bacteriol.,* 94, 403, 1967.
214. **Tachibana, D. K., Hayflick, L., and Rosenberg, L. T.,** Effect of complement and genetically defective mouse complement on *Mycoplasma canis, J. Infect. Dis.,* 121, 541, 1970.
215. **Lin, J.-S. and Kass, E. H.,** Immune inactivation of T-strain mycoplasmas, *J. Infect. Dis.,* 122, 93, 1970.
216. **Dörner, I., Brunner, H., Schiefer, H.-G., and Wellensiek, H.-J.,** Complement-mediated killing of *Acholeplasma laidlawii* by antibodies to various membrane components, *Infect. Immun.,* 13, 1663, 1976.
217. **Brunner, H., Dörner, I., Schiefer, H.-G., Krauss, H., and Wellensiek, H.-J.,** Lysis of *Acholeplasma laidlawii* by antibodies and complement, *Infect. Immun.,* 13, 1671, 1976.
218. **Dahl, J. S., Dahl, C. E., and Levine, R. P.,** Role of lipid fatty acyl composition and membrane fluidity in the resistance of *Acholeplasma laidlawii* to complement-mediated killing, *J. Immunol.,* 123, 104, 1979.
219. **Dörner, I., Brunner, H., Schiefer, H.-G., Loos, M., and Wellensiek, H.-J.,** Antibodies to *Acholeplasma laidlawii* membrane lipids in normal guinea pig serum, *Infect. Immun.,* 18, 1, 1977.
220. **Sussman, M., Jones, J. H., Almeida, J. D., and Lachmann, P. J.,** Deficiency of the second component of complement associated with anaphylactoid purpura and presence of mycoplasma in the serum, *Clin. Exp. Immunol.,* 14, 531, 1973.
221. **Lachmann, P. J.,** Genetic deficiencies of the complement system, *Boll. Ist. Sieroter. Milan.,* 53, 195, 1974.

222. **Bladen, H. A., Evans, R. T., and Mergenhagen, S. E.**, Lesions in *Escherichia coli* membranes after actions of antibody and complement, *J. Bacteriol.*, 91, 2377, 1966.
223. **Gewurz, H., Mergenhagen, S. E., Nowotny, A., and Phillips, J. K.**, Interactions of the complement system with native and chemically modified endotoxins, *J. Bacteriol.*, 95, 397, 1968.
224. **Snyderman, R., Gewurz, H., and Mergenhagen, S. E.**, Interaction of the complement system with endotoxic lipopolysaccharide. Generation of a factor chemotactic for polymorphonuclear leukocytes, *J. Exp. Med.*, 128, 259, 1968.
225. **Miler, I., Tlaskalova, H., Kostka, J., and Jilek, M.**, Effect of endotoxin on the complement level in sera of precolostral newborn pigs, *Folia Microbiol.*, 11, 475, 1966.
226. **Snyderman, R., Gewurz, H., Mergenhagen, S. E., and Jensen, J.**, Effect of C4 depletion on the utilization of the terminal components of guinea-pig complement by endotoxin, *Nature (London) New Biol.*, 231, 152, 1971.
227. **Muschel, L. H., Schmoker, K., and Webb, P. M.**, Anticomplementary action of endotoxin, *Proc. Soc. Exp. Biol. Med.*, 117, 639, 1964.
228. **Gewurz, H., Shin, H. S., and Mergenhagen, S. E.**, Interactions of the complement system with endotoxic lipopolysaccharide: consumption of each of the six terminal complement components, *J. Exp. Med.*, 128, 1049, 1968.
229. **Gewurz, H., Snyderman, R., Shin, H. S., Lichtenstein, L., and Mergenhagen, S. E.**, Complement (C') consumption by endotoxic lipopolysaccharide (LPS) in immunoglobulin deficient sera, *J. Clin. Invest.*, 47, 39a, 1968.
230. **Gewurz, H., Pickering, R. J., Snyderman, R., Lichtenstein, L. M., Good, R. A., and Mergenhagen, S. E.**, Interactions of the complement system with endotoxic lipopolysaccharides in immunoglobulin-deficient sera, *J. Exp. Med.*, 131, 817, 1970.
231. **Frank, M. M., May, J. E., and Kane, M. A.**, Contributions of the classical and alternate complement pathways to the biological effects of endotoxin, *J. Infect. Dis.*, 128, S176, 1973.
232. **Kane, M. A., May, J. E., and Frank, M. M.**, Interactions of the classical and alternate complement pathway with endotoxin lipopolysaccharide. Effect on platelets and blood coagulation, *J. Clin. Invest.*, 52, 370, 1973.
233. **Fine, D. P.**, Activation of the classic and alternate complement pathways by endotoxin, *J. Immunol.*, 112, 763, 1974.
234. **Füst, G., Bertok, L., and Juhasz-Nagy, S.**, Interactions of radio-detoxified *Escherichia coli* endotoxin preparations with the complement system, *Infect. Immun.*, 16, 26, 1977.
235. **Cooper, N. R. and Morrison, D. C.**, Binding and activation of the first component of human complement by the lipid A region of lipopolysaccharides, *J. Immunol.*, 120, 1862, 1978.
236. **Loos, M., Bitter-Suermann, D., and Dierich, M.**, Interaction of C$\bar{1}$, C2, and C4 with different preparations of bacterial lipopolysaccharides and with lipid A, *J. Immunol.*, 112, 935, 1974.
237. **Dierich, M. P., Bitter-Suermann, D., König, W., Hadding, U., Galanos, C., and Rietschel, E. T.**, Analysis of bypass activation of C3 by endotoxic LPS and loss of this potency, *Immunology*, 24, 721, 1973.
238. **Galanos, C., Rietschel, E. T., Lüderitz, O., and Westphal, O.**, Interaction of lipopolysaccharides and lipid A with complement, *Eur. J. Biochem.*, 19, 143, 1971.
239. **Morrison, D. C. and Kline, L. F.**, Activation of the classical and properdin pathways of complement by bacterial lipopolysaccharides (LPS), *J. Immunol.*, 118, 362, 1977.
240. **Spink, W. W., Davis, R. B., Potter, R., and Chartrand, S.**, The initial stage of canine endotoxin shock as an expression of anaphylactic shock: studies on complement titers and plasma histamine concentrations, *J. Clin. Invest.*, 43, 696, 1964.
241. **Aasen, A. O., Mellbye, O. J., and Ohlsson, K.**, Complement activation during subsequent stages of canine endotoxin shock, *Scand. J. Immunol.*, 8, 509, 1978.
242. **Füst, G. and Keresztes, M.**, Effect of endotoxin on the serum level of complement components. II. Effect of endotoxin on dog serum complement levels *in vivo* and *in vitro*, *Acta Microbiol. Acad. Sci. Hung.*, 16, 135, 1969.
243. **From, A. H. L., Gewurz, H., Gruninger, R. P., Pickering, R. J., and Spink, W. W.**, Complement in endotoxin shock: effect of complement depletion on the early hypotensive phase, *Infect. Immun.*, 2, 38, 1970.
244. **Kitzmiller, J. L., Lucas, W. E., and Yelenosky, P. F.**, The role of complement in feline endotoxin shock, *Am. J. Obstet. Gynecol.*, 112, 414, 1972.
245. **Gilbert, V. E. and Braude, A. I.**, Reduction of serum complement in rabbits after injection of endotoxin, *J. Exp. Med.*, 116, 477, 1962.
246. **Ulevitch, R. J., Cochrane, C. G., Henson, P. M., Morrison, D. C., and Doe, W. F.**, Mediation systems in bacterial lipopolysaccharide-induced hypotension and disseminated intravascular coagulation. I. The role of complement, *J. Exp. Med.*, 142, 1570, 1975.

247. **McCabe, W. R.**, Serum complement levels in bacteremia due to Gram-negative organisms, *N. Engl. J. Med.*, 288, 21, 1973.
248. **Füst, G., Petras, G., and Ujhelyi, E.**, Activation of the complement system during infections due to Gram-negative bacteria, *Clin. Immunol. Immunopathol.*, 5, 293, 1976.
249. **Fearon, D. T., Ruddy, S., Schur, P. H., and McCabe, W. R.**, Activation of the properdin pathway of complement in patients with Gram-negative bacteremia, *N. Engl. J. Med.*, 292, 937, 1975.
250. **Bessa, S. M., Dalmasso, A. P., and Goodale, R. L., Jr.**, Studies on the mechanism of endotoxin-induced increase of alveolocapillary permeability, *Proc. Soc. Exp. Biol. Med.*, 147, 701, 1974.
251. **Brown, D. L. and Lachmann, P. J.**, The behaviour of complement and platelets in lethal endotoxin shock in rabbits, *Int. Arch. Allergy Appl. Immunol.*, 45, 193, 1973.
252. **Müller-Berghaus, G. and Lohmann, E.**, The role of complement in endotoxin-induced disseminated intravascular coagulation: studies in congenitally C6-deficient rabbits, *Br. J. Haematol.*, 28, 403, 1974.
253. **Evensen, S. A., Pickering, R. J., Batbouta, J., and Shepro, D.**, Endothelial injury induced by bacterial endotoxin: effect of complement depletion, *Eur. J. Clin. Invest.*, 5, 463, 1975.
254. **Lichtenstein, L. M., Gewurz, H., Adkinson, N. F., Jr., Shin H. S., and Mergenhagen, S. E.**, Interactions of the complement system with endotoxic lipopolysaccharide: the generation of an anaphylatoxin, *Immunology*, 16, 327, 1969.
255. **Drake, W. P., Pokorney, D. R., Kopyta, L. P., and Mardiney, M. R., Jr.**, *In vivo* decomplementation of guinea pigs with cobra venom factor and anti-C3 serum: analysis of the requirement of C3 and C5 for the mediation of endotoxin-induced death, *Biomedicine*, 25, 91, 1976.
256. **Johnson, K. J. and Ward, P. A.**, The requirement for serum complement in the detoxification of bacterial endotoxin, *J. Immunol.*, 108, 611, 1972.
257. **Johnson, K. J. and Ward, P. A.**, Protective function of C6 in rabbits treated with bacterial endotoxin, *J. Immunol.*, 106, 1125, 1971.
258. **May, J. E., Kane, M. A., and Frank, M. M.**, Host defense against bacterial endotoxemia. Contribution of the early and late components of complement to detoxification, *J. Immunol.*, 109, 893, 1972.
259. **Bergstein, J. M. and Michael, A. F.**, Generalized Schwartzman reaction in the rabbit. Immunopathologic findings in the kidney, *Arch. Pathol.*, 97, 230, 1974.
260. **Stafford, B. T., Rapaport, S. I., and Shen, S. M.-C.**, Effects of endotoxin in cortisone-treated rabbits with a hereditary deficiency of the sixth component of complement (C6 deficiency), *Thromb. Haemostas.*, 35, 460, 1976.
261. **Bergstein, J. M. and Michael, A. F.**, Failure of cobra venom factor to prevent the generalized Schwartzman reaction and loss of renal cortical fibrinolytic activity, *Am. J. Pathol.*, 74, 19, 1974.
262. **Fong, J. S. C. and Good, R. A.**, Prevention of the localized and generalized Schwartzman reactions by an anticomplementary agent, cobra venom factor, *J. Exp. Med.*, 134, 642, 1971.
263. **Polak, L. and Turk, J. L.**, Suppression of the haemorrhagic component of the Schwartzmann reaction by anti-complement serum, *Nature (London)*, 223, 738, 1969.

Chapter 4

FUNGI

Fungal diseases are often considered prototypical of infections in which cell-mediated immunity plays a central role. Such an orientation may explain the relative paucity of investigations into the contribution of the complement system to host defense or organism pathogenicity. Nevertheless, a number of papers suggest a potential importance of complement.

Those fungi producing invasive destructive disease of deeper body structures (e.g., lung, brain, bone, liver, subcutaneous tissues) are called the "deep fungi"; their diseases, "systemic mycoses". The deep fungi are distinguished from the superficial fungi, which produce trivial or annoying but not life-threatening epidermal infections (e.g., athlete's foot). This chapter will be concerned only with deep fungal infections.

Fungi generally reproduce asexually by mitotic spore formation. They exist as yeasts (round structures whose spores form either externally as buds or internally as endospores) or molds (elongated structures called hyphae, which multiply linearly or in branches). The dimorphic fungi are yeast-like in tissue, hyphal on artificial media and in the proper natural environment. Table 1 lists some of the more important deep fungi.

The cell wall of fungi is composed of polysaccharides and chitin. *Cryptococcus neoformans* also contains an outer polysaccharide capsule.[1]

I. ASPERGILLUS

Aspergillus species are ubiquitous microorganisms which cause invasive disease primarily in immunosuppressed patients. In addition, aspergillus has been implicated as one of the agents in hypersensitivity lung disease. This syndrome affects patients with an atopic predisposition, who become heavily colonized in the bronchial tree with aspergillus. Precipitating serum antibodies to aspergillus extracts are demonstrable in high titers.[2] Complement has been detected by immunofluorescence in involved lungs, and some patients have had low serum complement levels during exacerbations.[3] Extracts of organisms associated with this syndrome, including *Aspergillus fumigatus*, *A. niger*, and *A. clavatus*, activated complement in vitro, as detected by immunoelectrophoretic conversion of factor B[3] and C3[4] as well as depression of CH50, C1, C4, C2, and C3.[5] The complement activation appeared to be primarily classical pathway.[5]

Budzko and Negroni[4] were able to dissociate the toxic and complement-fixing activities of the aspergillus extract. However, their measure of toxicity was the systemic reaction in mice and guinea pigs following intraperitoneal injection. Olenchock and Burrell[6] developed a more pertinent in vivo model of aerosolization. Inhalation of human serum albumin by rabbits immunized to albumin resulted in decline in arterial oxygen concentrations, a sensitive measure of pneumonitis. Nonimmunized rabbits given aerosols of *A. fumigatus* or *A. terreus* developed comparable hypoxemia and concomitant hypocomplementemia. The hypoxemia could be prevented by prior complement depletion with cobra venom factor. These investigators[6] questioned whether the reaction reflected natural or cross-reacting antibodies against aspergillus or some "nonspecific" activation of complement.

Table 1
SOME DEEP FUNGI ASSOCIATED WITH HUMAN DISEASES

Molds
    Aspergillus fumigatus and other species (aspergillosis)
    Rhizopus and *Mucor* species (phycomycosis)
Yeasts
    Candida albicans and other species (candidiasis)
    Torulopsis glabrata
    Cryptococcus neoformans (cryptococcosis)
Dimorphic fungi
    Histoplasma capsulatum (histoplasmosis)
    Blastomyces dermatitidis (North American blastomycosis)
    Sporothrix schenckii (sporotrichosis)
    Coccidioidis immitis (coccidioidomycosis)
    Paracoccidioides brasiliensis (South American blastomycocis)

II. CANDIDA

Species of candida are normal inhabitants of the skin and mucous membranes. They may cause mucocutaneous or systemic disease in otherwise healthy people but are more classically agents of disease in patients with diabetes mellitus or immunological impairment or in patients who have received multiple antibiotics. Candida may produce disseminated fungemia and infection of any organ system. In tissue, the organisms may appear in characteristic yeast form, in which buds may resemble hyphae ("pseudohyphae"), or as actual hyphae.

Morelli and Rosenberg[7] demonstrated a significantly shorter survival time of C5-deficient mice injected intravenously with 10^5–10^6 *Candida albicans*. The lethal doses intravenously or intraperitoneally however were not different from those in complement-sufficient mice. Gelfand and colleagues[8] found C4-deficient guinea pigs to be normal in response to systemic candidiasis. However, cobra venom factor-treated animals had a greater mortality rate, with increased numbers of yeasts in kidneys and spleen.

Sohnle and Kirkpatrick[9] produced experimental infections on shaved skin in guinea pigs. C3 was deposited along the basement membrane of the infected skin in normal and C4-deficient pigs. Immunoglobulin, C4, or candida antigen could not be detected. In a similar model Ray and Wuepper[10-12] observed that cobra venom factor treatment or C5 deficiency diminished the clinical severity of the lesions but allowed deeper tissue invasion by several species of candida.

Chronic mucocutaneous candidiasis is a disease associated with impaired cellular immunity to candida and characterized by chronic disfiguring hypertrophic candida infections. Skin biopsies in half the patients so studied had basement membrane deposition of properdin and C3 but not C4 or immunoglobulin.[13] The role of complement activation was only speculative.

A variety of strains of Candida have been shown to activate the complement system in vitro: *C. albicans*,[10,12,14] *C. tropicalis*,[12,14] *C. stellatoidea*,[12] *C. parapsilosis*,[12] *C. krusei*,[12] and *C. guilliermondi*.[12] When tested, complement activation has been through the alternative pathway, as evidenced by failure of EGTA to inhibit activation[10,14,15] or by normal activation in C4- or C2-deficient sera.[14,15] Kinetic studies comparing the classical and alternative pathways have not been done, however. The active factor from *C. albicans* appeared to be a mannan extracted from growth phase yeast and able to induce immunoelectrophoretic conversion of C3 and factor B in C4-deficient serum.[16]

Candida species have been shown to generate chemotactic factors from serum, probably via the alternative pathway.[10,14,15,17] Lehrer and Cline[18] observed a heat-labile serum requirement for phagocytosis of *C. albicans* by leukocytes. However, Davies and Denning[19] pointed out that one can often observe fungi growing in and out of leukocytes from patients or experimental animals with candidiasis. They showed that only a portion of *C. albicans* were killed by neutrophils and serum; the percentage was highly dependent on the inoculum size and ranged from 30 to 75%. In their studies, the required serum factor appeared to be complement and not antibody. Morelli and Rosenberg[20] confirmed the requirement for complement (which they thought might include C5 or other terminal components) but showed an enhancing effect of antiserum.

Lehrer and Cline[18] had not been able to show a requirement for any heat-labile, calcium-dependent factor in intracellular killing of *C. albicans* by human polymorphonuclear leukocytes. However, Yamamura and Valdimarsson[21] used an intricate assay system to distinguish ingestion and intracellular killing. In normal serum, yeast were ingested (uridine uptake by free *C. albicans* was inhibited) and killed (^{51}Cr was released from intracellular yeast). In heated serum, C2-deficient serum, or C3-deficient serum, killing but not uptake was diminished. The inhibition of killing in C2-deficient serum could be overcome by increasing the serum concentration. C6-deficient serum supported normal killing, as did C3-deficient serum with exogenous C3. The authors concluded that C3 was required for intracellular killing.

III. *TORULOPSIS GLABRATA*

This organism is microbiologically similar to candida and produces similar disease though less commonly. It is almost exclusively a pathogen of immunosuppressed patients. The organism induced a chemotactic factor upon incubation with fresh serum; heat-inactivated (56°C, 30 min) serum did not support this reaction.[17] Furthermore, heating to 50°C for 30 min (in order to inactivate factor B only) also partially inhibited the reaction.[22] This fact, plus the normal chemotactic factor activation in C2-deficient serum, suggested a role for the alternative pathway. Opsonization also appeared to be effective through the alternative pathway but quantitatively was only partially dependent on complement anyway.[22]

IV. *CRYPTOCOCCUS NEOFORMANS*

Cryptococcus neoformans is widely encountered in the environment and probably commonly produces infection. Only rarely does symptomatic disease ensue. Some overt disease can be readily attributed to immunosuppression, but at least half the patients encountered will be immunologically normal, as best one can tell. Principal sites of infection include meninges, lung, and occasionally skin and bones.[23]

Gadebusch[24] showed that injection of cryptococci into mice lowered serum properdin levels and that progressive infection resulted in further depletion of properdin. Subsequently, it was found that cobra venom factor-treated guinea pigs had a shortened survival time following experimental cryptococcal infection.[25] Supporting the notion that the alternative pathway may play an important role was the "normal" survival time in C4-deficient guinea pigs.[25]

Cryptococci as well as cryptococcal capsular polysaccharide activated serum complement in vitro,[24,26] primarily through the alternative pathway (as indicated by equivalent consumption of C3–9 in normal and C4-deficient guinea pig serum).[26] In contrast, when consumption of factor B, immunoelectrophoretic conversion of factor

B, and chemotactic factor generation were used as criteria, only the intact organism, encapsulated or nonencapsulated, could be shown to activate complement in general and the alternative pathway in particular.[27] These results suggested that the cell wall was responsible for alternative pathway activation, a conclusion in keeping with results from studies of bacteria.

Shahar et al.[28] found that in vivo phagocytosis of cryptococci by mouse leukocytes required exogenous rabbit antiserum. Only small cryptococci were actually ingested; larger organisms were surrounded by cuffs or "rosettes" of neutrophils and macrophages. In contrast, phagocytes from rabbits (naturally resistant to cryptococcal infection) formed such rosettes in vivo or in vitro in the absence of demonstrable antibody. A heat-labile serum factor was required for this reaction. Others have confirmed phagocytosis of cryptococci and its dependence on heat-labile serum factors.[25,26,29,30] Diamond and colleagues[29] demonstrated that ingestion was rapid and followed rapidly by intracellular killing; antibody seemed to play little role in the ingestion. Opsonization was effective in C2- and C4-deficient sera but slower; optimum kinetics required an intact classical pathway.[26]

The capsule appears to be important in opsonization of cryptococci in ways similar to the effect of bacterial capsules. Diamond et al.[29] noted that human neutrophils (but not macrophages) killed heavily encapsulated cryptococci more slowly than those with small capsules. Mitchell and Friedman[30] also noted a clear correlation between capsular size and ingestion: cryptococci with large capsules were poorly phagocytized (1.25 yeast/rat macrophage) compared to those with moderate (2.93 yeast/macrophage) or small (5.70 yeast/macrophage) capsules. Based on their observation that the capsule did not inhibit complement activation, Laxalt and Kozel[27] suggested that the capsule inhibited the interaction of bound opsonins with receptors on phagocytes. As noted in the chapter on Bacteria, a similar hypothesis has been advanced to explain the antiphagocytic effect of the staphylococcal capsule.[31] The alternative that less C3 is actually bound to an encapsulated cryptococcus has not, however, been adequately evaluated using more quantitative techniques.[32]

The major clinical presentation of cryptococcosis is meningitis. Diamond et al.[26] found spinal fluid not to be opsonic for cryptococci except in the presence of exogenous complement. Moreover, cryptococci isolated from spinal fluid did not have demonstrable surface complement.[26] They suggested[25] that complement probably was not important once central nervous system disease supervened but rather was critical in limiting the numbers of organisms at initial pulmonary sites or in the vascular system.

Neutrophils and macrophages from patients with cryptococcosis have not had demonstrable impairment of phagocytic or fungicidal capacities.[29] Macher and colleagues,[33] however, noted that cryptococcemia (i.e., disease associated with positive fungal blood cultures) carried a poor prognosis and measured opsonic activity in sera from such patients. Three patients with persistently positive blood cultures and disseminated cryptococcosis had almost no serum opsonic activity for cryptococci. This defect correlated with profound depletion of hemolytic complement and C3. On immunoelectrophoresis, factor B was in the converted form (Bb). The only patient with demonstrable antibody also had depression of C4. Two patients previously fungemic but recovering on therapy at time of study and three patients with localized disease all had normal opsonic and complement assays. Guinea pigs infected with cryptococci and fungemic or injected intravenously with large numbers of heat-killed cryptococci developed similar hypocomplementemia and opsonic deficiencies. Cryptococcal polysaccharide could not duplicate this effect. A radioimmunoassay for in vitro cryptococcal-bound human C3 demonstrated deposition of C3 in normal but not heat-inactivated serum; deposition was not inhibited by inhibition of the classical pathway (MgEGTA chelation, C2 deficiency) or absorption of antibody. Serum from the fungemic patients deposited little C3 on the cryptococci. The authors of this important paper suggested that fungemia led to a complement consumptive state that depleted serum opsonic activity and perhaps was causally related to the worsened prognosis.

V. *COCCIDIOIDES IMMITIS*

Coccidioides immitis, the agent of San Joaquin Valley fever, is environmentally confined to the southwestern U.S., Mexico, Central America, and parts of South America. As with other fungal diseases, exposure and infection are common among people in these areas, but disease is rare. The fungus exists in soil in hyphal (mycelial) form; spores developing in segments along a hypha may become airborne. If inhaled, the spores can produce primary pulmonary infection that may or may not develop into overt disease. The tissue phase is yeast-like (spherule), but daughter cells form as endospores.[34] Filtrates of mycelia and spherules generated chemotactic factors from normal human serum in vitro.[35] A heat-labile serum factor was required.

VI. *PARACOCCIDIOIDES BRASILIENSIS*

Paracoccidioides brasiliensis causes a granulomatous disease of skin, mucous membranes, lungs, and reticuloendothelial tissue. The fungus is localized to parts of Central and South America.[36] Calich et al.[37] demonstrated complement activation by and opsonization of yeast-phase organisms in human and guinea pig sera. For both effects, the alternative pathway seemed sufficient, though kinetic studies were not done. Immune sera were also opsonic, even if complement were inactivated. However, when neutrophil accumulation around fungi was studied in vivo, C3 depletion or C5 deficiency did not seem critical.

VII. SUMMARY AND COMMENTS

Though few investigators have examined the role of complement in mycoses, there have been rather uniform observations that fungi can activate the complement system, generally through the alternative pathway. Such activation generates chemotactic factors and leads to opsonization, especially of yeast forms. A role for complement in normal defenses has been most strongly suggested for candida species and *Cryptococcus neoformans*.

On the other hand, enthusiasm must be tempered by certain other observations. With the exception of candida, infection with most fungi is not characterized by polymorphonuclear leukocyte accumulation and is not noticeably more common in neutropenic patients. The hallmarks of deep fungal infections are chronicity and granulomatous inflammation. They tend to be associated with defects of cellular immunity.

Nevertheless, it seems a reasonable hypothesis that one of the principal early defenses against deep fungal disease is complement-mediated phagocytosis. Such a normal defense may partially explain the small proportion of infected persons who develop disease. The subject seems deserving of more study.

REFERENCES

1. **Bennett, J. E.,** Mycoses. Introduction, in *Principles and Practice of Infectious Diseases*, Mandell, G. L., Douglas, R. G., Jr., and Bennett, J. E., Eds., John Wiley & Sons, New York, 1979, p. 1979.
2. **Bennett, J. E.,** Aspergillus species, in *Principles and Practice of Infectious Diseases*, Mandell, G. L., Douglas, R. G., Jr., and Bennett, J. E., Eds., John Wiley & Sons, New York, 1979, 2002.
3. **Marx, J. J., Jr. and Flaherty, D. K.,** Activation of the complement sequence by extracts of bacteria and fungi associated with hypersensitivity pneumonitis, *J. Allergy Clin. Immunol.*, 57, 328, 1976.
4. **Budzko, D. B. and Negroni, R.,** Depletion of complement *in vivo* and *in vitro* by extracts of *Aspergillus fumigatus*, *Int. Arch. Allergy Appl. Immunol.*, 51, 518, 1976.
5. **DeBracco, M. M., Budzko, D. B., and Negroni, R.,** Mechanisms of activation of complement by extracts of *Aspergillus fumigatus*, *Clin. Immunol. Immunopathol.*, 5, 333, 1976.
6. **Olenchock, S. A. and Burrell, R.,** The role of precipitins and complement activation in the etiology of allergic lung disease, *J. Allergy Clin. Immunol.*, 58, 76, 1976.
7. **Morelli, R. and Rosenberg, L. T.,** Role of complement during experimental *Candida* infection in mice, *Infect. Immun.*, 3, 521, 1971.
8. **Gelfand, J. A., Hurley, D. L., Fauci, A. S., and Frank, M. M.,** Role of complement in host defense against experimental disseminated candidiasis, *J. Infect. Dis.*, 138, 9, 1978.
9. **Sohnle, P. G. and Kirkpatrick, C. H.,** Deposition of complement in the lesions of experimental cutaneous candidiasis in guinea pigs, *J. Cutan. Pathol.*, 3, 232, 1976.
10. **Ray, T. L. and Wuepper, K. D.,** Experimental cutaneous *Candida albicans* infections in rodents: role of complement, *Clin. Res.*, 23, 230A, 1975.
11. **Ray, T. L. and Wuepper, K. D.,** Experimental cutaneous candidiasis in rodents, *J. Invest. Dermatol.*, 66, 29, 1976.
12. **Ray, T. L. and Wuepper, K. D.,** Experimental cutaneous candidiasis in rodents. II. Role of the stratum corneum barrier and serum complement as a mediator of a protective inflammatory response, *Arch. Dermatol.*, 114, 539, 1978.
13. **Sohnle, P. G., Frank, M. M., and Kirkpatrick, C. H.,** Deposition of complement components in the cutaneous lesions of chronic mucocutaneous candidiasis, *Clin. Immunol. Immunopathol.*, 5, 340, 1976.
14. **Ray, T. L. and Wuepper, K. D.,** Activation of the alternative (properdin) pathway of complement by *Candida albicans* and related species, *J. Invest. Dermatol.*, 67, 700, 1976.
15. **Thong, Y. H. and Ferrante, A.,** Alternative pathway of complement activation by *Candida albicans*, *Aust. N.Z. J. Med.*, 8, 620, 1978.
16. **Ray, T. L., Hanson, A., Ray, L. F., and Wuepper, K. D.,** Purification of a mannan from *Candida albicans* which activates serum complement, *J. Invest. Dermatol.*, 73, 269, 1979.
17. **Denning, T. J. V. and Davies, R. R.,** *Candida albicans* and the chemotaxis of polymorphonuclear neutrophils, *Sabouraudia*, 11, 210, 1973.
18. **Lehrer, R. I. and Cline, M. J.,** Interaction of *Candida albicans* with human leukocytes and serum, *J. Bacteriol.*, 98, 996, 1969.
19. **Davies, R. R. and Denning, T. J. V.,** *Candida albicans* and the fungicidal activity of the blood, *Sabouraudia*, 10, 301, 1972.
20. **Morelli, R. and Rosenberg, L. T.,** The role of complement in the phagocytosis of *Candida albicans* by mouse peripheral blood leukocytes, *J. Immunol.*, 107, 476, 1971.
21. **Yamamura, M. and Valdimarsson, H.,** Participation of C3 in intracellular killing of *Candida albicans*, *Scand. J. Immunol.*, 6, 591, 1977.
22. **Ferrante, A. and Thong, Y. H.,** Activation of the alternative complement pathway by *Torulopsis glabrata*, *Scand. J. Infect. Dis.*, 11, 77, 1979.
23. **Diamond, R. D.,** *Cryptococcus neoformans*, in *Principles and Practice of Infectious Diseases*, Mandell, G. L., Douglas, R. G., Jr., and Bennett, J. E., Eds., John Wiley & Sons, New York, 1979, 2023.
24. **Gadebusch, H. H.,** Natural host resistance to infection with *Cryptococcus neoformans*. I. The effect of the properdin system on the experimental disease, *J. Infect. Dis.*, 109, 147, 1961.
25. **Diamond, R. D., May, J. E., Kane, M., Frank, M. M., and Bennett, J. E.,** The role of late complement components and the alternate complement pathway in experimental cryptococcosis, *Proc. Soc. Exp. Biol. Med.*, 144, 312, 1973.
26. **Diamond, R. D., May, J. E., Kane, M. A., Frank, M. M., and Bennett, J. E.,** The role of the classical and alternate complement pathways in host defenses against *Cryptococcus neoformans* infection, *J. Immunol.*, 112, 2260, 1974.

27. **Laxalt, K. A. and Kozel, T. R.**, Chemotaxigenesis and activation of the alternative complement pathway by encapsulated and nonencapsulated *Cryptococcus neoformans*, *Infect. Immun.*, 26, 435, 1979.
28. **Shahar, A., Kletter, Y., and Aronson, M.**, Granuloma formation in cryptococcosis, *Israel J. Med. Sci.*, 5, 1164, 1969.
29. **Diamond, R. D., Root, R. K., and Bennett, J. E.**, Factors influencing killing of *Cryptococcus neoformans* by human leukocytes *in vitro*, *J. Infect. Dis.*, 125, 367, 1972.
30. **Mitchell, T. G. and Friedman, L.**, *In vitro* phagocytosis and intracellular fate of variously encapsulated strains of *Cryptococcus neoformans*, *Infect. Immun.*, 5, 491, 1972.
31. **Wilkinson, B. J., Sisson, S. P., Kim, Y., and Peterson, P. K.**, Localization of the third component of complement on the cell wall of encapsulated *Staphylococcus aureus* M: implications for the mechanism of resistance to phagocytosis, *Infect. Immun.*, 26, 1159, 1979.
32. **Verbrugh, H. A., van Dijk, W. C., van Erne, M. E., Peters, R., Peterson, P. K., and Verhoef, J.**, Quantitation of the third component of human complement attached to the surface of opsonized bacteria: opsonin-deficient sera and phagocytosis-resistant strains, *Infect. Immun.*, 26, 808, 1979.
33. **Macher, A. M., Bennett, J. E., Gadek, J. E., and Frank, M. M.**, Complement depletion in cryptococcal sepsis, *J. Immunol.*, 120, 1686, 1978.
34. **Stevens, D. A.**, *Coccidioides immitis*, in *Principles and Practice of Infectious Diseases*, Mandell, G. L., Douglas, R. G., Jr., and Bennett, J. E., Eds., John Wiley & Sons, New York, 1979, 2053.
35. **Galgiani, J. N., Isenberg, R. A., and Stevens, D. A.**, Chemotaxigenic activity of extracts from the mycelial and spherule phases of *Coccidioides immitis* for human polymorphonuclear leukocytes, *Infect. Immun.*, 21, 862, 1978.
36. **Restrepo, M. A.**, *Paracoccidioides brasiliensis* (South American blastomycosis), in *Principles and Practice of Infectious Diseases*, Mandell, G. L., Douglas, R. G., Jr., and Bennett, J. E., Eds., John Wiley & Sons, New York, 1979, 2076.
37. **Calich, V. L. G., Kipnis, T. L., Mariano, M., Neto, C. F., and Dias da Silva, W.**, The activation of the complement system by *Paracoccidioides brasiliensis in vitro*: its opsonic effect and possible significance for an *in vivo* model of infection, *Clin. Immunol. Immunopathol.*, 12, 21, 1979.

Chapter 5

VIRUSES

Unlike bacteria, fungi, and protozoa, viruses are not fully self-contained, self-sufficient cells. They mainly consist of nucleic acids (RNA or DNA) surrounded by a protective protein coat. Also, unlike most of the microbes discussed in this text, viruses can multiply only within host cells. They are totally dependent on host synthetic and energy-generating systems for their replication.[1]

Viruses may infect animal, plant, and bacterial cells. Viruses that infect each of these groups of organisms are, in general, quite distinct, such that animal viruses are not capable of infecting plants and vice versa. However, information regarding structure and function of viruses may be generally applicable. Table 1 lists animal viruses that are pathogenic for man or are discussed in this chapter.

The core nucleic acid of a virus (the genome) and the surrounding protein shell (the capsid) comprise the nucleocapsid. The capsid is composed of repeating subunits that may be single proteins or more complex multiprotein subunits called capsomeres. This shell protects the genome from environmental hazards such as enzymes and also mediates introduction of the genome into the host cell. Capsids adsorb to cell membranes and then, through multiple penetration devices, inject the genome. The proteins of the capsid are virus specific, their synthesis being coded by viral nucleic acids. Other viral proteins may also be associated with the nucleic acid as internal or core proteins.[2]

The first step in virus interaction with the host cell is attachment, which may be mediated by ionic attraction and by specific receptors. The virus then penetrates the cell, following which the nucleic acid is separated from the protein capsid and incorporated into host nucleic acid. Uncoating of the genome may occur at the cell surface during penetration, in the cytoplasm, or at the nuclear membrane; the genome may be incorporated into the nucleus or may remain in the cytoplasm. The next step in the process is replication of the viral genome and stimulation of ribosomal protein synthesis with formation of new viral nucleic acid and proteins. New virions are assembled from new genomes and capsids. These progeny are then released in one of two ways: the host cell may disintegrate with release of new virus particles (a lytic infection); alternatively, viruses may bud from the cell surface, taking with them the surrounding coat (envelope) of host cell membrane.[3] The site for budding of these viruses appears to be at areas of cell membrane into which viral proteins have been incorporated during the synthetic phase of viral replication.[3] All but six families of animal viruses acquire such an envelope (Table 2).[2] Thus, these viruses have not only a capsid but also an envelope to protect the genome. For some viruses, all of the host proteins in the envelope are replaced by viral proteins; for others, the envelope contains both host and viral proteins.[2] The outer surface of the envelope may also contain virus-specific glycoprotein "spikes" demonstrable by electron microscopy.[2]

Inside the host cell, viruses utilize the cell's synthetic and energy systems for production of new viruses. The effects upon the host cell itself depend upon the number of viruses that may have gained access, the rate of viral replication or the extent to which host machinery is required by the virus, and perhaps the nature of the host cell itself. Viral replication interferes with normal cellular biosynthesis, partly by competition for cellular machinery; viral proteins may also inhibit host macromolecular synthesis. In addition, the cellular genome may be deregulated and

Table 1
VIRUSES OF MAN AND SOME OTHER ANIMALS[a]

DNA Viruses
  Poxviruses
    Variola major virus (smallpox)
    Variola minor virus (alastrim)
    Vaccinia virus
    Cowpox virus
    Orf virus (contagious pustular dermatitis)
    Molluscum contagiosum virus
  Herpesviruses
    Herpes simplex virus types 1 & 2
    Varicella-Zoster virus (chicken pox, herpes zoster)
    Cytomegalovirus
    Epstein-Barr virus (infectious mononucleosis)
    Infectious bovine rhinotracheitis virus
  Adenoviruses
    Human adenoviruses (31 serotypes)
  Papovaviruses
    Human papilloma viruses (warts)
    Polyoma virus
  Parvoviruses
    Aleutian mink disease virus
    Norwalk agent
RNA Viruses
  Picornaviruses
    Human enteroviruses
      Poliovirus
      Coxsackie viruses (A & B)
      ECHO virus
    Encephalomyocarditis virus
    Rhinoviruses
    Hepatitis A virus
  Togaviruses
    Alphaviruses
      Eastern equine encephalitis virus
      Semliki Forest virus
      Sindbis virus
      Chikungunya virus
      O'Nyong-Nyong virus
      Ross River virus
      Venezuelan encephalitis virus
      Western equine encephalitis virus
    Flaviviruses
      Yellow fever virus
      Dengue virus
      Japanese encephalitis virus
      St. Louis encephalitis virus
      Murray Valley encephalitis virus
      West Nile virus
      Kyasanur Forest virus
      Central European tickborne encephalitis virus
      Russian spring-summer encephalitis virus
      Louping ill virus
      Powassan virus
      Omsk hemorrhagic fever virus
    Rubella virus
    Equine arteritis virus

Table 1 (Continued)

    Bunyaviruses
        California encephalitis virus
        Rift Valley fever virus
        Crimean hemorrhagic fever virus
        Sandfly fever virus
    Reoviruses
        Colorado tick fever virus
        Rotaviruses
    Orthomyxoviruses
        Influenza viruses A, B, and C
    Paramyxoviruses
        Parainfluenza viruses types 1—5
        Newcastle disease virus
        Mumps virus
        Measles virus
        Respiratory syncytial virus
    Rhabdoviruses
        Vesicular stomatitis virus
        Rabies virus
        Marburg virus
        Ebola virus
    Retroviruses
    Arenaviruses
        Lymphocytic choriomeningitis virus
        Tacaribe virus complex
            Junin virus (Argentine hemorrhagic fever)
            Machupo virus (Bolivian hemorrhagic fever)
        Lassa virus
    Coronaviruses
        Infectious bronchitis virus
    Unclassified viruses
        Hepatitis B virus
        Hepatitis "non-A, non-B" virus(es)
        Chronic infectious neuropathic agents
            Kuru virus
            Jakob-Creutzfeldt virus
            Scrapie virus
            Transmissible mink encephalitis virus

[a] Adapted from Joklik.[2]

host synthetic genes newly expressed. New antigenic determinants either of viral or host specificity may appear on the surface of the infected cells. All these events may have either of two final effects on the host cell.[4] The interference with normal cell biosynthesis may induce abnormalities of plasma and lysosomal membranes, such abnormalities leading to permeability changes of the host cell and autolysis due to leakage of the lysosomes. The final result in this situation would be necrosis and lysis of the host cell. Alternatively, the host cell may be so altered or dedifferentiated as to assume primitive or even neoplastic characteristics.

    Essential features of viral replication and pathogenesis are that viruses, though they can remain active outside a cell, must gain entrance to a host cell in order to replicate. Released from one cell, they must then spread to nearby cells in order to further viral infection. It is during passage from one cell to another that viruses are most susceptible to destruction or neutralization. (Some viruses appear to spread directly from one cell to another without exposure to the extracellular environment

Table 2
CHARACTERISTICS OF VIRAL COATS

    Enveloped viruses
      Herpesviruses
      Togaviruses
      Bunyaviruses
      Orthomyxoviruses
      Paramyxoviruses
      Rhabdoviruses
      Retroviruses
      Arenaviruses
      Coronaviruses
    Naked viruses
      Adenoviruses
      Papovaviruses
      Parvoviruses
      Picornaviruses
      Reoviruses
      Poxviruses

and thus to avoid contact with humoral immune systems.) Host defense systems may also inhibit viral replication by destruction of virus-infected cells.[5]

The host immunological system is stimulated by and reactive with antigens on the surface of the capsid, on the surface of the envelope, or on the surface of the infected host cell.[6] Though viruses are exceedingly small, they express multiple antigens on these various surfaces with which the immune system may interact. Function of viral surface antigens (e.g., hemagglutinins) can be inhibited by some of the antibodies produced. Other antibodies activate complement in classical complement-fixation reactions. Finally, immunoglobulins may neutralize viruses directly. Neutralization may be studied either in vivo, in which case antibodies inhibit ability of the virus to produce disease in an experimental animal, or in vitro, in which case antibodies inhibit infectivity of viruses for cells in culture.[5] Neutralizing antibodies alone afford protection by inhibiting attachment, penetration, or uncoating.[7] With other effector systems, particularly complement and phagocytes, antibody neutralization may be mediated through viral lysis, phagocytosis of the virus-antibody complex, or destruction of the virus-infected host cell.[8]

In addition to lysis of the virus or the virus-infected cell, complement may participate in defenses against viral infections in at least three other ways (Table 3). Antibody alone, though attached to viruses, may not be fully capable of neutralization. In such circumstances, addition of complement components, primarily components of the classical pathway and C3, may enhance or complete neutralization.[9] This enhancing effect is thought to be due to further protein accumulation on the virus surface. Complement may also agglutinate infectious virus-antibody complexes, decreasing the effective number of infectious particles and inhibiting virus function.[7] Finally, complement alone neutralizes some virus in the absence of antibody.[7]

Oldstone[7] has emphasized that immune complexes formed with virus and antibody, especially in the presence of complement, not only protect the host but also mediate the immunopathology of many viral infections. Viral antigens, antibody, and complement have been deposited in glomeruli and other blood vessels in a variety of acute and chronic viral infections.[7,10] Porter[8] has emphasized that anti-

Table 3
DEMONSTRATED MECHANISMS OF COMPLEMENT-MEDIATED VIRAL NEUTRALIZATION

Surface accretion of host proteins
    Vaccinia virus (13, 17)[a]
    Herpes simplex virus (28—30, 32)
    Equine arteritis virus (77—79)
    Newcastle disease virus (85)
Viral aggregation
    Herpes simplex virus (31)
    Polyoma virus (57)
    Lymphocytic choriomeningitis virus (107, 108)
Viral lysis
    Sindbis virus (63)
    Equine arteritis virus (78, 79)
    Newcastle disease virus (86)
    Retroviruses (97, 100—106)
    Lymphocytic choriomeningitis virus (108)
Host cell lysis
    Vaccinia virus (14—16)
    Herpes simplex virus (14, 33—37)
    Infectious bovine rhinotracheitis virus (55, 56)[b]
    Sindbis virus (64)
    Semliki Forest virus (64)
    Influenza A virus (36, 82)
    Measles virus (36, 88—92)
    Rabies virus (95)
    Retroviruses (98)

[a] References in parentheses.
[b] Requires polymorphonuclear leukocytes or lymphocytes.

body-complement mediated cytolysis of virus-infected cells may also, in addition to containing infection, mediate symptoms of disease.

Despite evidence of activity of the humoral immune system, some viral infections, particularly those due to the herpes-viruses, appear to be mainly under the control of the cellular immune system.

I. POXVIRUSES

A. Variola Virus (Smallpox)

Thompson[11] measured serial complement activity in 15 patients with smallpox in 1903. He used killing of typhoid bacilli as an assay for complement. Complement values (i.e., serum bactericidal activity) were low early in infections but rapidly returned to normal in mild cases. The four patients who developed secondary (generally streptococcal) infections maintained lower levels of bactericidal activity until recovery or death.

B. Vaccinia Virus

One of the earliest studies on in vitro interactions of viruses and the complement system was performed in 1930 by Douglas and Smith[12] using vaccinia virus. Incubation of virus with normal blood blocked the usual cutaneous reaction to subsequent intradermal injection in rabbits. Activity was dependent on cells of the blood, since plasma alone was not inhibitory. However, serum or plasma was also required and the activity in plasma was heat labile. Immunity enhanced inhibitory activity of the blood; the immune serum factor was not heat labile.

Gipson et al.[13] demonstrated that complement enhanced neutralization of vaccinia virus by IgG antibody. Rheumatoid factor further enhanced complement-mediated neutralization. Complement also mediated lysis of vaccinia virus-infected cells in tissue culture.[14-16].

Complement-dependent antibody-mediated neutralization of vaccinia virus infection of tissue culture cells was normal in sera deficient in C5 and C6.[17] However, C1r-deficient serum showed only slight neutralizing activity and C4-deficient serum was no better than heat-inactivated serum. Both C2- and C3-deficient sera showed definite but subnormal neutralizing activity. The requirement for the complement components C1, C4, C2, and C3 could be overcome in this system by increasing the concentration of antibody; this result suggested that the complement components were most important early in infections when antibody titers were low. Because of the clear-cut dose effect, these authors[17] attributed the effect of complement to accumulation of "extraneous protein" on the viral envelope.

II. HERPES VIRUSES

A. Herpes Simplex Virus

Herpes simplex virus causes recurrent superficial ulcerations of mucous membranes and a devastating form of meningoencephalitis. Strunk et al.[18] studied the course of infection in C4-deficient guinea pigs. The virus, introduced intradermally, produced similar skin lesions in normal and C4-deficient guinea pigs and was cleared similarly from the skin. Both groups of animals developed comparable antibody titers and levels of C1 and C3–9. Likewise, intraperitoneal infection resulted in similar clinical responses: clearance of viruses from peritoneum, spleen, and blood; antibody responses; and complement levels.

However, neutralizing antibody formed in response to herpes simplex infection in humans was at least partially complement dependent.[19,20] And the majority of sera obtained during early stages of herpes simplex encephalitis contained primarily complement-dependent neutralizing antibody.[21] The neutralizing antibody response (IgG and IgM) in experimental animals immunized with herpes simplex virus showed a similar early complement dependence with a switch to complement-independence later or in hyperimmune states.[22-26]

The steps of sensitization by antibody and subsequent neutralization by complement have been dissected.[27] Viruses incubated with antibody in vitro were slowly and inefficiently neutralized over 3 hr. Addition of complement either to the original mixture of virus and antibody or at any subsequent time markedly shortened the interval to neutralization and increased the number of viral particles neutralized. Using purified complement components, Daniels et al.[28] demonstrated that C4, in high concentration, neutralized virus-IgM-C1 complexes; addition of C2 and C3 did not further enhance neutralization. At lower C4 concentrations, C2 and C3 enhanced neutralization. These results were further substantiated by observations that C4-deficient serum would not neutralize herpes simplex virus-IgM complexes, whereas sera deficient in C5 or C6 were neutralizing.[29] Daniels et al.[29] attributed these effects of the complement system to the "piling up" of proteins on the surface of the virus. Supporting that interpretation were demonstrations that rheumatoid factor further enhanced neutralization by complement.[30]

Wallis and Melnick[31] suggested that complement may also enhance neutralization by aggregation of herpes virus particles. Incubation of virus-antibody complexes with complement decreased the number of infectious units as well as increased the size of viral particles. Yoshino and Kishie[32] disagreed with this notion and suggested that aggregation could not explain neutralization of herpes virus by complement. They based this argument on the observation that the effect of complement was not related to the concentration of the virus. However, the two mechanisms of viral neutralization, accretion of surface proteins and aggregation, are not necessarily mutually exclusive and both may be involved.

Complement also mediates lysis of host cells infected with herpes simplex virus.[14,33-35] Perrin and colleagues[36] found that such lysis could be mediated by the alternative complement pathway. IgG was required but activity appeared to reside in the F(ab')$_2$ fragment. (Their studies did not rule out a role for the classical pathway as well, and by immunofluorescence they demonstrated C4 and C2 on the virus surface.) In this regard it is interesting to consider observations by Tompkins et al.[37] that not all infected cells were equally susceptible to lysis by antibody and complement. Susceptibility of otherwise resistant cells was enhanced by treatment with neuraminidase. It will be remembered from Chapter 1 that similar treatment renders a variety of cells susceptible to alternative pathway lysis.

Austin and Daniels[38,39] demonstrated that staphylococcal protein A could inhibit complement-mediated neutralization. They attributed this effect to protein-A-mediated complement activation in serum and to blockage of complement-activating sites on immunoglobulins bound to viruses or virus-infected cells. They postulated that this observation may underlie the clinically recognized association of staphyloccal and viral infections.

A factor released in vitro from herpes simplex virus-infected cells has been shown to activate complement and generate a chemotactic factor, probably C5a.[40] The factor was elaborated within 12 hr of infection and elaboration closely parallelled onset of cytopathology. The authors hypothesized that viral-induced lysis released a factor which could generate chemotactic activity from complement.

A family with recurrent oral herpes simplex infections has been described by Kapadia et al.[41] At times of active infection, the proband had markedly decreased hemolytic titers of whole complement, C5, and factor B as well as a modestly depressed level of C1. Other components (C4, C2, C6, C7, C9, and C3) were normal. At a later observation, when C3 was normal by immunochemical measurement, hemolytic titers of C3 were decreased. Those family members with history of recurrent herpes infections had low levels of at least one of the later acting components at some time during observation and most had low levels of factor B. The propositus also had a history of recurrent otitis media, pharyngitis, and arthralgias. These observations might have suggested a genetic propensity to herpes infections and perhaps to other infections. However, the patient's husband also had low complement levels associated with recurrent herpes simplex virus infections. Shared exposure to an unusual strain of herpes simplex virus may have produced this familial illness. Thus, the herpes infections may have induced complement deficiency and thereby propensity to bacterial infections. Other patients studied have not had changes in complement levels with recurrent mucosal herpes simplex infections.[19]

B. Cytomegalovirus

Cytomegalovirus infections range from asymptomatic to a syndrome resembling mononucleosis. In immunosuppressed patients, an overwhelming infection may occur. Neutralizing antibody formed early following immunization of rabbits with murine cytomegalovirus was almost exclusively complement dependent. Later neutralizing activity was largely complement independent.[42] One patient had modestly low levels of C3 and C4 during acute infection with cytomegalovirus but no clinical evidence of immune-complex disease.[43] However, circulating complement-fixing immune complexes have been demonstrated in patients with congenitally acquired cytomegalovirus infection.[44] And an adult with cytomegalovirus pneumonia developed nephritis associated with glomerular deposition of immunoglobulins, complement, and cytomegaloviral antigen.[45]

C. Epstein-Barr Virus

This virus is the etiologic agent of the majority of cases of infectious mononucleosis and has been implicated in the etiology of Burkitt's lymphoma and nasopharyngeal carcinoma. Many of the clinical features of infectious mononucleosis are compatible with immune-complex disease. Wands and colleagues[46] reported a case of infectious mononucleosis characterized by urticaria in addition to more classical features of adenopathy, pharyngitis, and hepatosplenomegaly. During the urticarial phase the patient had circulating cryoproteins containing particles similar to the Epstein-Barr virus, antibodies to Epstein-Barr virus, IgM, IgG, IgA, and complement components C4, C3, and C5. Charlesworth and colleagues[47] studied 34 patients with Epstein-Barr virus mononucleosis. Of 31 patients with uncomplicated disease, 10 had low levels of C4. Of these ten, six had evidence of C3 conversion products in fresh plasma and eight had at least one other component abnormally low. Three of the 34 patients had a course complicated by hemolytic anemia, arthralgias and myalgias, or nephritis. All of these patients had low levels of C3 and C3 conversion products in plasma; two of the three also had low levels of C1q or C4, C5, and C7.

Lymphoblastoid cells, derived from Burkitt's lymphoma tissue or from peripheral mononuclear cells of patients with infectious mononucleosis, activated the alternative complement pathway in human, rat, and mouse serum but not in guinea pig serum.[48] Activation was independent of antibody and could, in chicken serum, result in lysis of the cells. Those lymphoblastoid cells with surface Epstein-Barr virus antigen were much more effective than those without the surface antigen in activation of the alternative pathway.[49] Such activation may be a normal host defense system.

The Epstein-Barr virus generally infects only cells of the B lymphocyte system. In vitro, the virus binds to B lymphocytes but not T lymphocytes. One of the characteristic features of B lymphocytes is a receptor for C3. Jondal and colleagues[50] demonstrated that receptors for the Epstein-Barr virus and for C3 are closely linked on human B lymphocytes. Both receptors were expressed on the same cells, attachment of the virus inhibited subsequent attachment of complement-coated particles, the two receptors capped together, and lymphocytes lacking complement receptors could not be infected with Epstein-Barr virus. The receptors did not appear to be identical, since blockage of the complement receptor did not inhibit virus binding; furthermore, the virus did not inhibit complement receptors on macrophages. Nevertheless, the receptors were probably on the same molecule, since increasing the amount of protein (anti-C3 followed by anti-immunoglobulin) at the C3 receptor could block virus binding. These observations have been confirmed with other methods and other B cell lines as well as with human peripheral blood B lymphocytes.[51,52] The expression of the Epstein-Barr virus and complement receptors did not correlate however with the ability of cell lines to activate the complement system.[53] This curious association of the two receptors presumably reflects a viral adaptation for attachment and penetration of host cells.[50]

D. Infectious Bovine Rhinotracheitis Virus

Rossi and Kiesel[54] studied kinetics of production of complement-dependent neutralizing antibodies following vaccination of cattle with this virus. A single injection induced both IgM and IgG responses; neutralization by both antibody classes was complement dependent. A second injection of vaccine enhanced the concentration of IgG neutralizing antibody, but this antibody was only partially complement dependent.

Bovine rhinotracheitis virus has been used to study the phenomenon of antibody-dependent cell-mediated cytotoxicity, the lysis of target cells by certain Fc-receptor bearing cells (see Chapter 1). Antibody mediates attachment of the "killer" cell to the target cell. However, an Fc receptor on the "killer" cell, while necessary, may not always be sufficient for the reaction. Rouse et al.[55] demonstrated enhancement of cytolysis by complement at various effector/target ratios, at various antibody concentrations (including concentrations at which antibody alone no longer mediated cytotoxicity), and at various incubation times. Never was antibody alone, even with prolonged incubation, equivalent to the combination of antibody and complement. These studies were performed with polymorphonuclear leukocytes as effector cells. In this system of virus-infected target cells, lymphocytes do not ordinarily mediate cytotoxicity but can be induced to do so by incorporation of complement into the reaction. The authors attributed the effect of complement to increased binding between effector and target cells and suggested that antibody-dependent cell-mediated cytotoxicity may represent an early host defense against viral infection. In such circumstances, when antibody is limited, complement may be critical in enhancing bonding and thus killing.[56]

III. PAPOVAVIRUSES

Polyoma virus is a potentially oncogenic DNA virus that can cause chronic viral infections associated with immune-complex formation. Oldstone et al.[57] demonstrated that complement enhanced neutralization of the virus by antibody. The terminal components C5–9 were not required; the early components of the classical pathway C1, 4, 2, and C3 were required. In contrast to herpes simplex virus, no neutralizing effect could be demonstrated until C3 was added. The authors suggested that the effect of the complement components was primarily mediated by agglutination of the virus-antibody complexes rather than by lysis or accretion of protein.

IV. PICORNAVIRUSES

A. Enteroviruses

Yuceoglu et al.[58] described two children with acute glomerulonephritis and profound hypocomplementemia probably related to infection with ECHO virus type 9. The organism was isolated from the stool of one of the patients and both developed neutralizing antibody to ECHO virus type 9. Evidence for other etiologies was absent. Subsequently Burch et al.[59] induced glomerulonephritis in mice by injection of ECHO virus type 9, however complement studies were not performed.

B. Hepatitis A Virus

Immune-complex disease is a well-recognized feature of hepatitis B infection (see below). However, only a few examples of this phenomenon with hepatitis A have been published. Fernandez and McCarty[60] presented three cases with viral hepatitis associated with a prodrome of arthritis, rash, petechiae, and urticaria. All these patients lacked evidence for hepatitis B infection though hepatitis A was not documented. (Other hepatitis viruses ["non-A, non-B"] cause similar disease but have yet to be identified.) The only patient on whom complement studies were performed during acute illness had a normal CH50. More recently Baer et al.[61] reported 50 patients studied during an outbreak of hepatitis A. Of eight children tested, seven had low levels of C3 during acute illness.

V. TOGAVIRUSES

A. Alphaviruses

The alphaviruses were previously classified as group A arboviruses. In development of neutralization tests for antibody to Semliki Forest virus and Sindbis virus, it was demonstrated that a factor in normal serum enhanced neutralization by immune guinea pig serum.[62] This factor had certain characteristics in common with complement (heat lability, divalent cation dependency, and inhibition by other methods often used to inhibit complement). Stollar[63] demonstrated that antibody and complement could induce direct lytic damage to Sindbis virus with release of viral RNA. This reaction is in addition to the more usual antibody-complement-mediated cytolysis of host cells infected with Sindbis or Semliki Forest viruses.[64]

Hirsch et al.[65] studied Sindbis virus infection in newborn mice, which develop an acute and often fatal encephalitis. Mice treated with cobra venom factor developed a disease that was similar but characterized by greater concentrations of virus in blood and brain and a more prolonged course. The overall mortality was similar, but death occurred later in the cobra venom factor-treated animals. The brains of the complement-depleted animals demonstrated more extensive inflammation. The authors suggested that complement had a dual role in this infection, limiting viremia but also mediating some of the adverse inflammatory reaction.

Some of the early studies of the role of complement in virus neutralization were performed with Western equine encephalitis virus. Morgan[66] noted that the neutralization titer of serum was diminished upon prolonged standing at 4°C or by heating and that activity could be restored by complement (fresh guinea pig serum). Neutralization by partially purified rabbit antibody was also enhanced by complement. Whitman[67] demonstrated a similar phenomenon in convalescent human serum. Dozois et al.[68] further analyzed the factor in normal serum that enhanced the virus-neutralizing activity of immune serum. Using then-available methods to inactivate various components, they suggested that complement (specifically C2, C3, and C4) was indeed the enhancing factor. C1 appeared to be required only in small amounts, if at all.

B. Flaviviruses

Flaviviruses were previously classified as group B arboviruses. Of the group, only dengue virus has been evaluated regarding interactions with complement. Sabin[69] demonstrated that neutralization of the virus was dependent on heat-stable specific antibody and a heat-labile nonspecific plasma factor.

The dengue shock syndrome is characterized by several days of fever, followed by rapidly progressive hypotension due to increased vascular permeability and intravascular volume depletion. The disease is characteristically seen in children who have serological evidence of previous infection with another dengue serotype. It has been suggested that the syndrome is immunologically mediated and represents an immune-complex disease. Russell and colleagues[70] studied 12 children with the dengue shock syndrome. C3 levels were moderately to markedly depressed during the shock phase and returned to normal during convalescence. All of these children had serologic evidence of previous dengue infection. Other patients studied during primary infection sometimes had low normal complement levels. These investigators suggested that the low complement values reflected the immunopathology of the syndrome. Bokisch et al.[71] confirmed these observations using more sensitive complement assays. All of the patients with dengue shock syndrome had decreased levels of all complement components except C9 during the phase of shock. Furthermore the degree of shock correlated with the degree of complement deficiency. Confirming the complement activation were studies of increased C3 and C1q catabolism.[71-73] Patients with the shock syndrome also had evidence of disseminated intravascular coagulation with decreased fibrinogen levels, circulating fibrin split products, and thrombocytopenia.[71] In subsequent studies, hypocomplementemia and thrombocytopenia have generally correlated with shock and coagulopathy.[74,75] Though causal relationships among these disturbances have not been established, activation of the complement system is a prominent part of the dengue shock syndrome.

C. Rubella Virus

Rubella virus infection of cells in tissue culture can be inhibited (neutralized) by antibody; as with many other viruses, this inhibition is enhanced by heat-labile serum components.[76] This enhancing activity appeared to be complement because of its heat lability and removal from serum by immune complexes. Unlike most of the viruses studied, the complement dependency was seen with late- as well as early-developing antibodies.

D. Equine Arteritis Virus

Radwan and colleagues[77-79] have studied the complement requirements of neutralizing antibody to equine arteritis virus. Late-developing antibodies were also complement dependent. They separated stages of neutralization into binding of antibody to the virus, which remained infective, and subsequent addition of complement, which produced immediate neutralization.[77] They demonstrated also that complement was capable of lysing the virus as indicated by release of viral RNA.[78] However, neutralization (i.e., diminished infectivity) could be demonstrated earlier than lysis. These results suggested that neutralization was due to "steric hindrance" of "critical sites" on the viral envelope. Studies with purified complement components[79] demonstrated that the components C1-3 were sufficient for neutralization and addition of terminal components resulted in lysis. In the presence of adequate amounts of antibody or the first four components of complement, terminal components did not enhance neutralization. However, when antibody and C1, C4, C2, and C3 were limited, C5-9 enhanced neutralization. The authors suggested that neutralization by antibody and the first four components of complement was most important, but early in infection, when antibody was present in low concentration, the lytic effect of complement might be important.[79]

VI. ORTHOMYXOVIRUSES

Influenza viruses are major world-wide pathogens of man and other animals. Most serious infections in man are produced by influenza A and influenza B. These viruses possess characteristic surface antigens that serve viral functions; these are the hemagglutinins and neuraminidase antigens. Major protective antibodies are directed against these antigens. The periodicity and virulence of infections with influenza are related, especially in influenza A, to the ability of these antigens to change gradually (antigenic drift) or dramatically (antigenic shift), leaving the host defense systems unprepared for the antigenically new virus. Ginsburg and Horsfall[80] demonstrated in 1949 that in vitro neutralization and hemagglutination inhibition by normal serum from humans, guinea pigs, and rabbits depended on a factor that was heat labile, lost on storage at 4°C, and calcium dependent. Based on the best available methodology, the authors concluded that this factor was not complement; however, in retrospect it seems likely they were indeed measuring the contribution of complement. Styk[81] demonstrated that immunization with influenza A evoked an early antibody response that was largely dependent on a heat-labile factor in normal serum for neutralizing ability; neutralization by antibody formed later was largely heat stable. Virus-infected cells also express the hemagglutinin and may be lysed by antihemagglutinins and complement.[82] Perrin et al.[36] also demonstrated alternative pathway-mediated lysis of influenza-A-infected cell lines.

Hicks et al.[83] studied influenza A infection in normal mice, mice depleted of C3–9 with cobra venom factor, and C5-deficient mice. Both groups of complement-deficient mice had prolonged infection, increased pulmonary consolidation, and increased mortality. Antibody production was normal but virus was not cleared in cobra venom factor-treated animals until C3 levels recovered. Administration of cobra venom factor at different times related to infection demonstrated that the critical event was a normal C3 level at the usual time of virus clearance (7 to 9 days after infection).

VII. PARAMYXOVIRUSES

A. Newcastle Disease Virus

Wedgwood et al.[84] studied the interactions of Newcastle disease virus with the properdin system. Agglutination of erythrocytes and infectivity for chick embryo yolk sacs were inhibited by heat-labile serum factors. Inhibition was impaired in serum depleted of properdin and activity could be restored by addition of properdin. The four then-known components of complement were also required. Subsequently Linscott and Levinson[85] confirmed that neutralization of Newcastle disease virus by IgM antibody formed early in the immune response required the contributions of C1, 4, 2, and 3. These investigators[85] felt that the major effect was "contributing bulk in the form of large protein molecules to the virus-antibody complex."

Virus-induced hemagglutination can be followed by lysis of the agglutinated erythrocytes. Apostolov and Sawa[86] demonstrated that pretreatment of the virus with antiserum plus complement enhanced subsequent hemolysis. They suggested that complement-mediated disruption of the viral envelope basement membrane facilitated fusion of the envelope with the erythrocyte membrane and subsequent lysis of the erythrocyte.

B. Measles Virus

Measles virus, in addition to producing the common disease of childhood, is related etiologically to subacute sclerosing panencephalitis and possibly to multiple sclerosis and partial lipodystrophy. Charlesworth and colleagues[87] measured complement titers during uncomplicated measles infection. Thirteen of the 50 patients had evidence of classical pathway activation with low levels of C1q, C4, and C3. These patients generally also had evidence by immunoelectrophoresis of C3 conversion. Twelve patients had low levels of C1q but no other abnormalities. And seven patients had evidence of alternative pathway activation with low levels of C3, evidence of C3 conversion, and normal levels of C1q and C4.

The clinical syndrome of measles may itself be due to a host immune response to virus-infected cells and not to direct viral damage. Joseph et al.[88] demonstrated that measles virus-antibody complexes activated the classical complement pathway, whereas virus-infected cells in the presence of immune human serum activated the alternative complement pathway. This activity was due to IgG antibodies. Perrin et al.[36] subsequently demonstrated that the activity of the immunoglobulin resided in the F (ab')$_2$ fragment. Hicks et al.[89] confirmed this alternative pathway activation, but suggested that the cytolytic system is most effective when both pathways are intact. In experimental infections the cytolytic, complement-dependent antibody appeared very early and may represent an important primary host defense system.[90] Such a function would not exclude a role in immunopathology of the disease.

The cytolytic antibodies induced by measles infection are not homogenous. Rather they are directed against a variety of antigens on the viral surface, most prominently the hemagglutinin and the hemolysin. Ehrnst[91,92] demonstrated that an IgG antihemagglutinin mediated cytotoxicity primarily through the alternative complement pathway and independently of the Fc fragment. In contrast, an IgG antihemolysin activated the classical complement pathway in an Fc-dependent fashion.

VIII. RHABDOVIRUSES

Mills and colleagues[93,94] demonstrated that human complement could neutralize vesicular stomatitis virus by a mechanism apparently independent of antibody. The reaction required the components of the classical pathway and C3, but C5–8 and factor B were not required. The activating factor appeared to be an envelope glycoprotein that bound C1q and (in the presence of an additional, unidentified serum factor) activated C1.

Tissue culture cells infected with rabies virus were shown to be lysed in the presence of antiserum and complement.[95] Evidence for the involvement of complement came from the ability of EDTA chelation, heat inactivation, or hydrazine to abolish the effect of serum.

IX. RETROVIRUSES

Retroviruses (oncornaviruses) are RNA tumor viruses of animals other than man and may be considered as a group. In 1931 Mueller[96] studied the effect of normal guinea pig serum on neutralization of Rous fowl sarcoma virus by chicken antiserum. A heat-labile factor in the guinea pig serum enhanced the protective effect of the immune serum. It was subsequently demonstrated that, in the presence of complement, an antibody to envelope proteins of the AKR mouse leukemia virus could produce lysis of the virus itself.[97] And antibodies have been identified that will, in a complement-dependent fashion, lyse host cells infected with feline lymphoma virus.[98] Interestingly, however, in at least one system, murine lymphoma cells were transformed by infection with Moloney leukemia virus such that the cells became resistant to complement-mediated lysis even though complement had been activated on the cell surface and typical complement "holes" could be demonstrated by electron microscopy.[99]

Retroviruses do not infect humans; this fact has been presumed evidence for a natural defense mechanism against such viruses in man. Welsh et al.[100] found human serum, in the absence of demonstrable specific antibodies, to lyse a variety of retroviruses. In contrast, guinea pig, rabbit, and mouse sera did not neutralize the viruses. Activity in human serum was heat labile and dependent on C2 and C4 but not antibody. All human sera tested have had this lytic activity for all retroviruses studied.[101,102] Subsequent investigations indicated that the reaction occurred directly between a virus surface protein and human C1q.[103] Resultant activation of C1s and the rest of the complement sequence led to lysis of the virus. Bartholomew and colleagues[104] subsequently identified the surface protein of Moloney leukemia virus responsible for activation of C1 as the membrane protein p15(E). This protein reacted directly with human and guinea pig C1q, but the binding to guinea pig C1q did not lead to activation of guinea pig C1s.[105] By recombination of the C1 molecule with human or guinea pig subunits, they demonstrated that the guinea pig C1s was the subunit that could not be activated; C1 molecules containing guinea pig C1q and C1r but human C1s could be activated by p15(E) with resultant viral lysis. Sherwin et al.[106] have questioned however whether this nonspecific lysis of retroviruses is an important defense system. If this postulate were true, they suggested, only sera of resistant animals should contain such activity. They found however that all retroviruses studied could be lysed by a variety of primate sera, including sera from animals susceptible to virus infection. There was no correlation between degree of lytic activity and ease of virus transmission in any species.

X. ARENAVIRUSES

Mice experimentally infected *in utero* or at birth with lymphocytic choriomeningitis virus develop a chronic degenerative disease with evidence of immune-complex damage to glomeruli and other vessels. Oldstone and Dixon[107] studied this infection in animals depleted of complement with cobra venom factor. Complement depletion, though it had no effect on the numbers of organisms isolated from the brains of infected animals, aborted the tissue injury associated with the disease. The LD50 in cobra venom factor-treated mice was over two logs higher than in normal mice. When cobra venom factor was discontinued and hemolytic complement titers returned to normal, protection rapidly waned and tissue injury ensued. Direct inhibitory effect of complement on the virus has also been demonstrated in vitro.[107,108] Antibody effected modest neutralization of the virus by aggregation; complement markedly enhanced this neutralization. Welsh et al.[108] also demonstrated that complement could induce lysis of the virus via the classical pathway.

Thus, with this virus, as with others, one can demonstrate two quite different but not necessarily contradictory effects of complement. Complement can destroy or neutralize the virus in vitro; however, in vivo complement primarily mediates clinical disease. Presumably both phenomena occur during actual infection. It must be remembered that neutralization experiments involve incubation of virus with antibody and complement for some period of time before attempted infection. This manipulation is probably quite different from situations that obtain during actual infection and favors host defense systems in the serum.

Interactions with the complement system may also be affected by changes in the virus induced by the host. Lymphocytic choriomeningitis virus and many others acquire surface antigens from host cells in which they replicate; antibody and complement directed against these antigens may also participate in damage to the virus.[109] Complement-fixing antibodies may also be directed against a variety of other antigens including core antigens of the virus,[110] though the role of these antibodies remains speculative.

Several arenaviruses cause devastating hemorrhagic fever syndromes. Hallmarks of these diseases include diffuse vasculitis, hemorrhage, and disturbances of the coagulation system. DeBracco and colleagues[111] demonstrated depressed levels of hemolytic complement, C2, C3, and C5 early in the course of Argentine hemorrhagic fever (Junin virus). Complement activity returned to normal about the same time as clinical recovery and about the time antibody could first be demonstrated. Low values correlated with severe disease. In patients with Thai hemorrhagic fever,[112] levels of C3 clearly correlated with severity of illness. Patients with profound shock and massive hemorrhage had very low complement values. There was no correlation between complement or severity of disease and evidence of disseminated intravascular coagulation.

XI. CORONAVIRUSES

Berry and Almeida[113] looked closely at the effect of various antisera on the avian infectious bronchitis virus, an enveloped virus. The virus was grown in chick embryo yolk sacs; antisera were raised in either chickens or rabbits. When chicken antiserum was incubated with virus, antibody could be demonstrated by electron microscopy only on viral projections. Presence or absence of complement made no difference in such deposition though fresh serum was more efficient in neutralization than heat-inactivated serum. When rabbit antiserum was incubated with virus, antibody could be demonstrated not only on projections but also on the envelope itself. In the presence of complement such antiserum induced typical complement "holes" in the viral envelope membrane. These results implied close antigenic relationships between the viral envelope and the host membrane from which the virus budded. These results also emphasized the importance of careful assessment of methodology in determining the importance of interactions between complement and various antisera.

XII. HEPATITIS B VIRUS

The virus of hepatitis B has not yet been fully characterized and for the time being is classified separate from other viruses, although its nucleic acid has been identified as DNA. The disease produced by this virus was previously called serum hepatitis in recognition of the classical mode of transmission. With ability to identify the viral antigen in blood, it has become clear that there are other modes of transmission as well and the disease cannot always be distinguished on clinical grounds from other forms of hepatitis. The principal antigen of the virus is called the Hepatitis B surface Antigen (HBsAg), previously referred to as hepatitis-associated antigen (HAA) or Australia antigen (AuAg). Overt hepatitis may be preceded by a prodrome resembling acute serum sickness, with arthritis or arthralgias, fever, or skin rashes.

Gocke and colleagues[114,115] found a strong association between polyarteritis nodosa (a chronic vasculitis) and chronic circulation of immune complexes containing HBsAg. Biopsies of involved vessels showed deposits of the antigen, immunoglobulins, and complement components. Furthermore, some of the patients had low complement values. Subsequent studies have confirmed the high association of HBsAg immune complexes and polyarteritis nodosa, though not all patients with polyarteritis have demonstrable hepatitis B antigenemia.[116]

There is an even stronger association between the prodrome of hepatitis and immune complexes containing hepatitis antigen.[116-119] Complement has been found in these complexes and patients have had low serum complement levels. Wands et al.[119] demonstrated complement components, immunoglobulins, and HBsAg in cryoprecipitates from serum of patients with arthritis and hepatitis. Patients with arthritis tended to have higher concentrations of cryoprecipitates, and their cryoprecipitates were more likely to contain complement and complement-activating subclasses of IgG. Alternative pathway involvement was suggested by presence of factor B activation products (Bb) in acute sera; furthermore, immune complexes in vitro could cleave factor B in fresh normal serum.

Low C3 values were also noted in a group of patients with acute hepatitis B without obvious immune-complex disease.[120] In a larger group of such patients,[121] however, only minimal decreases in component levels occurred; furthermore, antigen-negative patients with acute hepatitis and patients with cirrhosis (not due to viral hepatitis) had similarly diminished complement activity. It is not clear that patients with acute hepatitis actually have complement activation in the absence of overt immune-complex disease; low complement values, if they occur, may reflect decreased synthesis due to liver disease.

Chronic active hepatitis is associated in some but not all cases with hepatitis B virus infection and is characterized by destructive fibrosing hepatitis leading to cirrhosis. Of ten such patients studied,[122] those with extrahepatic manifestations had serum cryoproteins containing IgG specific for hepatitis B antigen. The subset of these patients with arthritis (as opposed to only arthralgias) also had IgM, IgA, and complement (C4, C3, C5) in cryoprecipitates. Three patients, during the active phase of arthritis, had low serum values of C3 and C4 and evidence of conversion of factor B. When C3 metabolism was evaluated in another group of such patients, the metabolic rate was diminished, suggesting that any low complement values were due to liver injury and not to complement activation.[123] However, since it is not clear how many of the patients in the second study had extrahepatic manifestations, the two studies are probably not strictly comparable.

Eknoyan et al.[124] evaluated seven patients with acute viral hepatitis, four of whom had circulating HBsAg, proteinuria, and diminished creatinine clearance. All seven had glomerulonephritis with focal deposits of immunoglobulins and C3. Kneiser et al.[125] reported three other cases of immune-complex-mediated nephritis during infection with hepatitis B virus. Their patients had a variety of histological pictures of nephritis ranging from membranous to interstitial to membranoproliferative, but all were associated with deposition of complement in the kidney. Chronic nephritides have also been described with chronic hepatitis B antigenemia[126] and chronic active hepatitis;[127] in both situations, complement has seemed to play a critical role. In one case[126] the additional presence of cryoprecipitates and clinical manifestations such as urticarial rash and arthralgias provided evidence of a link between nephritis and the other extrahepatic manifestations of hepatitis B. A strong association exists among Japanese children between membranous glomerulonephritis and chronic hepatitis B antigenemia.[128,129] Of 163 children with proteinuria or hematuria or both, 11 had membranous glomerulonephritis. All of these 11 had circulating HBsAg (this incidence compares with an incidence of 4.6% in the general childhood population). Six clearly had C3 deposited in the glomerulus.

XIII. BACTERIOPHAGES

Bacteriophages are viruses that are infective for bacteria. Investigators have used these viruses as models for the role of complement in virus neutralization. Normal human serum neutralized bacteriophages in a system requiring both antibody and complement.[130,131] Antibody produced in experimental animals in the early period following immunization was similar to this "normal" antibody in that neutralization was largely complement dependent.[132] Antibody produced later following immunization neutralized virus independently of complement.[132,133]

REFERENCES

1. **Joklik, W. K.**, The nature, isolation, and measurement of animal viruses, in *Zinsser Microbiology*, 17th ed., Joklik, W. K., Willett, H. P., and Amos, D. B., Eds., Appleton-Century-Crofts, New York, 1980, 966.
2. **Joklik, W. K.**, The structure, components, and classification of viruses, in *Zinsser Microbiology*, 17th ed., Joklik, W. K., Willett, H. P., and Amos, D. B., Eds., Appleton-Century-Crofts, New York, 1980, 980.
3. **Joklik, W. K.**, The virus multiplication cycle, in *Zinsser Microbiology*, 17th ed., Joklik, W. K., Willett, H. P., and Amos, D. B., Eds., Appleton-Century-Crofts, New York, 1980, 1040.
4. **Joklik, W. K.**, The effect of virus infection on the host cell, in *Zinsser Microbiology*, 17th ed., Joklik, W. K., Willett, H. P., and Amos, D. B., Eds., Appleton-Century-Crofts, New York, 1980, 1086.
5. **Merigan, T. C.**, Host defenses against viral disease, *N. Engl. J. Med.*, 290, 323, 1974.
6. **Joklik, W. K.**, Viruses and viral proteins as antigens, in *Zinsser Microbiology*, 17th ed., Joklik, W. K., Willett, H. P., and Amos, D. B., Eds., Appleton-Century-Crofts, New York, 1980, 1030.
7. **Oldstone, M. B. A.**, Virus neutralization and virus-induced immune complex disease. Virus-antibody union resulting in immunoprotection or immunologic injury — two sides of the same coin, *Progr. Med. Virol.*, 19, 84, 1975.
8. **Porter, D. D.**, Destruction of virus-infected cells by immunological mechanisms, *Annu. Rev. Microbiol.*, 25, 283, 1971.
9. **Notkins, A. L.**, Infectious virus-antibody complexes: interaction with anti-immunoglobulins, complement, and rheumatoid factor, *J. Exp. Med.*, 134, 41S, 1971.
10. **Oldstone, M. B. A. and Dixon, F. J.**, Immune complex disease in chronic viral infections, *J. Exp. Med.*, 134, 32S, 1971.
11. **Thompson, R. L.**, The bacteriolytic complement content of the blood serum in variola, *J. Med. Res.*, 10, 71, 1903.
12. **Douglas, S. R. and Smith, W.**, A study of vaccinal immunity in rabbits by means of *in vitro* methods, *Br. J. Exp. Pathol.*, 11, 96, 1930.
13. **Gipson, T. G., Daniels, C. A., and Notkins, A. L.**, Interaction of rheumatoid factor with infectious vaccinia virus-antibody complexes, *J. Immunol.*, 112, 2087, 1974.
14. **Brier, A. M., Wohlenberg, C., Rosenthal, J., Mage, M., and Notkins, A. L.**, Inhibition or enhancement of immunological injury of virus-infected cells, *Proc. Natl. Acad. Sci.*, 68, 3073, 1971.
15. **Hayashi, K., Lodmell, D., Rosenthal, J., and Notkins, A. L.**, Binding of ^{125}I-labeled anti-IgG, rheumatoid factor and anti-C3 to immune complexes on the surface of virus-infected cells, *J. Immunol.*, 110, 316, 1973.
16. **Gipson, T. G. and Daniels, C. A.**, The effect of rheumatoid factor on the immune lysis of vaccinia virus-infected cells, *Clin. Immunol. Immunopathol.*, 4, 16, 1975.
17. **Leddy, J. P., Simons, R. L., and Douglas, R. G.**, Effect of selective complement deficiency on the rate of neutralization of enveloped viruses by human sera, *J. Immunol.*, 118, 28, 1977.
18. **Strunk, R. C., John, T. J., and Sieber, O. F.**, Herpes simplex virus infection in guinea pigs deficient in the fourth component of complement. *Infect. Immun.*, 15, 165, 1977.
19. **Heineman, H. S.**, Herpes simplex neutralizing antibody — quantitation of the complement-dependent fraction in different phases of adult human infection, *J. Immunol.*, 99, 214, 1967.
20. **Schmidt, N. J., Forghani, B., and Lennette, E. H.**, Type specificity of complement-requiring and immunoglobulin M neutralizing antibody in initial herpes simplex virus infections of humans, *Infect. Immun.*, 12, 728, 1975.
21. **Lerner, A. M., Bailey, E. J., and Nolan, D. C.**, Complement-requiring neutralizing antibodies in *Herpesvirus hominis* encephalitis, *J. Immunol.*, 104, 607, 1970.
22. **Yoshino, K. and Taniguchi, S.**, Studies on the neutralization of herpes simplex virus. I. Appearance of neutralizing antibodies having different grades of complement requirement, *Virology*, 26, 44, 1965.

23. **Taniguchi, S. and Yoshino, K.,** Studies on the neutralization of herpes simplex virus. II. Analysis of complement as the antibody-potentiating factor, *Virology*, 26, 54, 1965.
24. **Hampar, B., Notkins, A. L., Mage, M., and Keehn, M. A.,** Heterogeneity in the properties of 7S and 19S rabbit-neutralizing antibodies to herpes simplex virus, *J. Immunol.*, 100, 586, 1968.
25. **Shinkai, K. and Yoshino, K.,** Complement requirement of neutralizing antibodies in different classes of immunoglobulin appearing in rabbits and guinea pigs after primary and booster immunizations with herpes simplex virus, *Jpn. J. Microbiol.*, 19, 25, 1975.
26. **Yoshino, K. and Isono, N.,** Studies on the neutralization of herpes simplex virus. IX. Variance in complement requirement among IgG and IgM from early and late sera under different sensitization conditions, *Microbiol. Immunol.*, 22, 403, 1978.
27. **Yoshino, K. and Taniguchi, S.,** Studies on the neutralization of herpes simplex virus. III. Mechanism of the antibody-potentiating action of complement, *Virology*, 26, 61, 1965.
28. **Daniels, C. A., Borsos, T., Rapp, H. J., Snyderman, R., and Notkins, A. L.,** Neutralization of sensitized virus by the fourth component of complement, *Science*, 165, 508, 1969.
29. **Daniels, C. A., Borsos, T., Rapp, H. J., Snyderman, R., and Notkins, A. L.,** Neutralization of sensitized virus by purified components of complement, *Proc. Natl. Acad. Sci.*, 65, 528, 1970.
30. **Ashe, W. K., Daniels, C. A., Scott, G. S., and Notkins, A. L.,** Interaction of rheumatoid factor with infectious herpes simplex virus-antibody complexes, *Science*, 172, 176, 1971.
31. **Wallis, C. and Melnick, J. L.,** Herpesvirus neutralization: the role of complement, *J. Immunol.*, 107, 1235, 1971.
32. **Yoshino, K. and Kishie, T.,** Studies on the neutralization of herpes simplex virus. VI. The mode of action of complement upon antibody-sensitized virus, *Jpn. J. Microbiol.*, 17, 63, 1973.
33. **Rager-Zisman, B. and Bloom, B. R.,** Immunological destruction of herpes simplex virus I infected cells, *Nature (London)*, 251, 542, 1974.
34. **McClung, H., Seth, P., and Rawls, W. E.,** Quantitation of antibodies to herpes simplex virus types 1 and 2 by complement-dependent antibody lysis of infected cells, *Am. J. Epidemiol.*, 104, 181, 1976.
35. **Shivers, J. C. and Daniels, C. A.,** Effect of rheumatoid factor and anti-immunoglobulin G antibodies on complement-mediated lysis of herpes simplex virus-infected human fibroblasts, *Infect. Immun.*, 15, 478, 1977.
36. **Perrin, L. H., Joseph, B. S., Cooper, N. R., and Oldstone, M. B. A.,** Mechanism of injury of virus-infected cells by antiviral antibody and complement: participation of IgG, F(ab')2, and the alternative complement pathway, *J. Exp. Med.*, 143, 1027, 1976.
37. **Tompkins, W. A. F., Seth, P., Gee, S., and Rawls, W. E.,** Neuraminidase reversal of resistance to lysis of herpes simplex virus-infected cells by antibody and complement, *J. Immunol.*, 116, 489, 1976.
38. **Austin, R. M. and Daniels, C. A.,** Interaction of staphylococcal protein A with virus-IgG complexes, *J. Immunol.*, 113, 1568, 1974.
39. **Austin, R. M. and Daniels, C. A.,** Effect of staphylococcal protein A on complement-potentiated neutralization of herpes simplex virus and immune lysis of virus-infected cells, *Infect. Immun.*, 12, 821, 1975.
40. **Brier, A. M., Snyderman, R., Mergenhagen, S. E., and Notkins, A. L.,** Inflammation and herpes simplex virus: release of a chemotaxis-generating factor from infected cells, *Science*, 170, 1104, 1970.
41. **Kapadia, A., Gupta, S., Good, R. A., and Day, N. K.,** Familial herpes simplex infection associated with activation of the complement system, *Am. J. Med.*, 67, 122, 1979.
42. **Kim, K. S. and Carp, R. I.,** Influence of complement on the neutralization of murine cytomegalovirus by rabbit antibody, *J. Virol.*, 12, 1620, 1973.
43. **Strunk, R. C., Sieber, O. F., Taussig, L. M., and Gall, E. P.,** Serum complement depression during viral lower respiratory tract illness in cystic fibrosis, *Arch. Dis. Child.*, 52, 687, 1977.
44. **Stagno, S., Volanakis, J. E., Reynolds, D. W., Stroud, R., and Alford, C. A.,** Immune complexes in congenital and natal cytomegalovirus infections of man, *J. Clin. Invest.*, 60, 838, 1977.
45. **Ozawa, T. and Stewart, J. A.,** Immune-complex glomerulonephritis associated with cytomegalovirus infection, *Am. J. Clin. Pathol.*, 72, 103, 1979.
46. **Wands, J. R., Perrotto, J. L., and Isselbacher, K. J.,** Circulating immune complexes and complement sequence activation in infectious mononucleosis, *Am. J. Med.*, 60, 269, 1976.
47. **Charlesworth, J. A., Pussell, B. A., Roy, L. P., Lawrence, S., and Robertson, M. R.,** The complement system in infectious mononucleosis, *Aust. N.Z. J. Med.*, 7, 23, 1977.
48. **Budzko, D. B., Lachmann, P. J., and McConnell, I.,** Activation of the alternative complement pathway by lymphoblastoid cell lines derived from patients with Burkitt's lymphoma and infectious mononucleosis, *Cell. Immunol.*, 22, 98, 1976.

49. **McConnell, I., Klein, G., Lint, T. F., and Lachmann, P. J.,** Activation of the alternative complement pathway by human B cell lymphoma lines is associated with Epstein-Barr virus transformation of the cells, *Eur. J. Immunol.*, 8, 453, 1978.
50. **Jondal, M., Klein, G., Oldstone, M. B. A., Bokisch, V., and Yefenof, E.,** Surface markers on human B and T lymphocytes. VIII. Association between complement and Epstein-Barr virus receptors on human lymphoid cells, *Scand. J. Immunol.*, 5, 401, 1976.
51. **Yefenof, E. and Klein, G.,** Membrane receptor stripping confirms the association between EBV receptors and complement receptors on the surface of human B lymphoma lines, *Int. J. Cancer*, 20, 347, 1977.
52. **Einhorn, L., Steinitz, M., Yefenof, E., Ernberg, I., Bakacs, T., and Klein, G.,** Epstein-Barr virus (EBV) receptors, complement receptors, and EBV infectibility of different lymphocyte fractions of human peripheral blood. II. Epstein-Barr virus studies, *Cell Immunol.*, 35, 43, 1978.
53. **Yefenof, E., Klein, G., and Kvarnung, K.,** Relationships between complement activation, complement binding, and EBV absorption by human hematopoietic cell lines, *Cell Immunol.*, 31, 225, 1977.
54. **Rossi, C. R. and Kiesel, G. K.,** Antibody class and complement requirement of neutralizing antibodies in the primary and secondary antibody response of cattle to infectious bovine rhinotracheitis virus vaccine, *Arch. Virol.*, 51, 191, 1976.
55. **Rouse, B. T., Grewal, A. S., and Babiuk, L. A.,** Complement enhances antiviral antibody-dependent cell cytotoxicity, *Nature (London)*, 266, 456, 1977.
56. **Rouse, B. T., Grewal, A. S., Babiuk, L. A., and Fujimiya, Y.,** Enhancement of antibody-dependent cell-mediated cytotoxicity of herpesvirus-infected cells by complement, *Infect. Immun.*, 18, 660, 1977.
57. **Oldstone, M. B. A., Cooper, N. R., and Larson, D. L.,** Formation and biologic role of polyoma virus-antibody complexes. A critical role for complement, *J. Exp. Med.*, 140, 549, 1974.
58. **Yuceoglu, A. M., Berkovich, S., and Minkowitz, S.,** Acute glomerulonephritis associated with ECHO virus type 9 infection, *J. Pediatr.*, 69, 603, 1966.
59. **Burch, G. E., Chu, K. C., and Sohal, R. S.,** Glomerulonephritis induced in mice by ECHO 9 virus, *N. Engl. J. Med.*, 279, 1420, 1968.
60. **Fernandez, R. and McCarty, D. J.,** The arthritis of viral hepatitis, *Ann. Intern. Med.*, 74, 207, 1971.
61. **Baer, G. M., Walker, J. A., and Yager, P. A.,** Studies of an outbreak of acute hepatitis A. I. Complement level fluctuation, *J. Med. Virol.*, 1, 1, 1977.
62. **Way, H. J. and Garwes, D. J.,** Serum accessory factors in the measurement of arbovirus neutralization reactions, *J. Gen. Virol.*, 7, 211, 1970.
63. **Stollar, V.,** Immune lysis of Sindbis virus, *Virology*, 66, 620, 1975.
64. **King, B., Wust, C. J., and Brown, A.,** Antibody-dependent, complement-mediated homologous and cross-cytolysis of togavirus-infected cells, *J. Immunol.*, 119, 1289, 1977.
65. **Hirsch, R. L., Griffin, D. E., and Winkelstein, J. A.,** The effect of complement depletion on the course of Sindbis virus infection in mice, *J. Immunol.*, 121, 1276, 1978.
66. **Morgan, I. M.,** Quantitative study of the neutralization of Western equine encephalomyelitis virus by its antiserum and the effect of complement, *J. Immunol.*, 50, 359, 1945.
67. **Whitman, L.,** The neutralization of Western equine encephalomyelitis virus by human convalescent serum. The influence of heat labile substances in serum on the neutralization index, *J. Immunol.*, 56, 97, 1947.
68. **Dozois, T. F., Wagner, J. C., Chemerda, C. M., and Andrew, V. M.,** The influence of certain serum factors on the neutralization of Western equine encephalomyelitis virus, *J. Immunol.*, 62, 319, 1949.
69. **Sabin, A. B.,** The dengue group of viruses and its family relationships, *Bacteriol. Rev.*, 14, 225, 1950.
70. **Russell, P. K., Intavivat, A., and Kanchanapilant, S.,** Antidengue immunoglobulins and serum β1 c/a globulin levels in dengue shock syndrome, *J. Immunol.*, 102, 412, 1969.
71. **Bokisch, V. A., Top, F. H., Jr., Russell, P. K., Dixon, F. J., and Müller-Eberhard, H. J.,** The potential pathogenic role of complement in dengue hemorrhagic shock syndrome, *N. Engl. J. Med.*, 289, 996, 1973.
72. **Bokisch, V. A., Müller-Eberhard, H. J., and Dixon, F. J.,** Complement — a potential mediator of the hemorrhagic shock syndrome (dengue), *Adv. Biosci.*, 12, 417, 1973.
73. **Bokisch, V. A.,** The role of complement in hemorrhagic shock syndrome (dengue), in *Progress in Immunology II, Vol. 4: Clinical Aspects I,* Brent, L., and Holborow, J., Eds., North-Holland, Amsterdam, 1974, 151.
74. **Edelman, R., Nimmannitya, S., Colman, R. W., Talamo, R. C., and Top, F. H., Jr.,** Evaluation of the plasma kinin system in dengue hemorrhagic fever, *J. Lab. Clin. Med.*, 86, 410, 1975.

75. **Phanichyakarn, P., Pongpanich, B., Israngkura, P. B., Dhanamitta, S., and Valyasevi, A.,** Studies on dengue hemorrhagic fever. III. Serum complement (C3) and platelet studies, *J. Med. Assoc. Thailand*, 60, 301, 1977.
76. **Rawls, W. E., Desmyter, J., and Melnick, J. L.,** Rubella virus neutralization by plaque reduction, *Proc. Soc. Exp. Biol. Med.*, 124, 167, 1967.
77. **Radwan, A. I. and Burger, D.,** The complement-requiring neutralization of equine arteritis virus by late antisera, *Virology*, 51, 71, 1973.
78. **Radwan, A. I., Burger, D., and Davis, W. C.,** The fate of sensitized equine arteritis virus following neutralization by complement or anti-IgG serum, *Virology*, 53, 372, 1973.
79. **Radwan, A. I. and Crawford, T. B.,** The mechanisms of neutralization of sensitized equine arteritis virus by complement components, *J. Gen. Virol.*, 25, 229, 1974.
80. **Ginsberg, H. S. and Horsfall, F. L., Jr.,** A labile component of normal serum which combines with various viruses. Neutralization of infectivity and inhibition of hemagglutination by the component, *J. Exp. Med.*, 90, 475, 1949.
81. **Styk, B.,** Cofactor and specific antibodies against influenza viruses. III. The potentiating effect of cofactor on specific antibodies of early immune and of hyperimmune sera and the differences in the character of these antibodies, *Acta Virol.*, 6, 327, 1962.
82. **Verbonitz, M. W., Enis, F. A., Hicks, J. T., and Albrecht, P.,** Hemagglutinin-specific complement-dependent cytolytic antibody response to influenza infection, *J. Exp. Med.*, 147, 265, 1978.
83. **Hicks, J. T., Ennis, F. A., Kim, E., and Verbonitz, M.,** The importance of an intact complement pathway in recovery from a primary viral infection: influenza in decomplemented and in C5-deficient mice, *J. Immunol.*, 121, 1437, 1978.
84. **Wedgwood, R. J., Ginsberg, H. S., and Pillemer, L.,** The properdin system and immunity. VI. The inactivation of Newcastle disease virus by the properdin system, *J. Exp. Med.*, 104, 707, 1956.
85. **Linscott, W. D. and Levinson, W. E.,** Complement components required for virus neutralization by early immunoglobulin antibody, *Proc. Natl. Acad. Sci.*, 64, 520, 1969.
86. **Apostolov, K. and Sawa, M. I.,** Enhancement of haemolysis by Newcastle disease virus (NDV) after pretreatment with heterophile antibody and complement, *J. Gen. Virol.*, 33, 459, 1976.
87. **Charlesworth, J. A., Pussell, B. A., Roy, L. P., Robertson, M. R., and Beveridge, J.,** Measles infection. Involvement of the complement system, *Clin. Exp. Immunol.*, 24, 401, 1976.
88. **Joseph, B. S., Cooper, N. R., and Oldstone, M. B. A.,** Immunologic injury of cultured cells infected with measles virus. I. Role of IgG antibody and the alternative complement pathway, *J. Exp. Med.*, 141, 761, 1975.
89. **Hicks, J. T., Klutch, M. J., Albrecht, P., and Frank, M. M.,** Analysis of complement-dependent antibody-mediated lysis of target cells acutely infected with measles, *J. Immunol.*, 117, 208, 1976.
90. **Hicks, J. T. and Albrecht, P.,** Cytolytic, complement-dependent antibodies to measles virus in rhesus monkeys after administration of live or killed virus, *J. Infect. Dis.*, 133, 648, 1976.
91. **Ehrnst, A. C.,** Complement activation by measles virus cytotoxic antibodies: alternative pathway C activation by hemagglutination-inhibition antibodies but classical activation by hemolysin antibodies, *J. Immunol.*, 118, 533, 1977.
92. **Ehrnst, A.,** Separate pathways of complement activation by measles virus cytotoxic antibodies: subclass analysis and capacity of F(ab) molecules to activate C via the alternative pathway, *J. Immunol.*, 121, 1206, 1978.
93. **Mills, B. J. and Cooper, N. R.,** Antibody-independent neutralization of vesicular stomatitis virus by human complement. I. Complement requirements, *J. Immunol.*, 121, 1549, 1978.
94. **Mills, B. J., Beebe, D. P., and Cooper, N. R.,** Antibody-independent neutralization of vesicular stomatitis virus by human complement. II. Formation of VSV-liproprotein complexes in human serum and complement-dependent viral lysis, *J. Immunol.*, 123, 2518, 1979.
95. **Wiktor, T. J., Kuwert, E., and Koprowski, H.,** Immune lysis of rabies virus-infected cells, *J. Immunol.*, 101, 1271, 1968.
96. **Mueller, J. H.,** The effect of alexin in virus-antivirus mixtures, *J. Immunol.*, 20, 17, 1931.
97. **Oroszlan, S. and Gilden, R. V.,** Immune virolysis: effect of antibody and complement on C-type RNA virus, *Science*, 168, 1478, 1970.
98. **Grant, C. K., Essex, M., Pedersen, N. C., Hardy, W. D., Jr., Stephenson, J. R., Cotter, S. M., and Theilen, G. H.,** Lysis of feline lymphoma cells by complement-dependent antibodies in feline leukemia virus contact cats. Correlation of lysis and antibodies to feline oncornavirus-associated cell membrane antigen, *J. Natl. Cancer Inst.*, 60, 161, 1978.
99. **Cooper, N. R., Polley, M. J., and Oldstone, M. B. A.,** Failure of terminal complement components to induce lysis of Moloney virus transformed lymphocytes, *J. Immunol.*, 112, 866, 1974.
100. **Welsh, R. M., Jr., Cooper, N. R., Jensen, F. C., and Oldstone, M. B. A.,** Human serum lyses RNA tumour viruses, *Nature (London)*, 257, 612, 1975.

101. **Welsh, R. M., Jr., Jensen, F. C., Cooper, N. R., and Oldstone, M. B. A.,** Inactivation and lysis of oncornaviruses by human serum, *Virology*, 74, 432, 1976.
102. **Jensen, F. C., Welsh, R. M., Cooper, N. R., and Oldstone, M. B. A.,** Lysis of oncornaviruses by human serum, *Bibl. Haematol.*, 43, 438, 1976.
103. **Cooper, N. R., Jensen, F. C., Welsh, R. M., Jr., and Oldstone, M. B. A.,** Lysis of RNA tumor viruses by human serum: direct antibody-independent triggering of the classical complement pathway, *J. Exp. Med.*, 144, 970, 1976.
104. **Bartholomew, R. M., Esser, A. F., and Müller-Eberhard, H. J.,** Lysis of oncornaviruses by human serum. Isolation of the viral complement (C1) receptor and identification as p15E, *J. Exp. Med.*, 147, 844, 1978.
105. **Bartholomew, R. M. and Esser, A. F.,** Differences in activation of human and guinea pig complement by retroviruses, *J. Immunol.*, 121, 1748, 1978.
106. **Sherwin, S. A., Benveniste, R. E., and Todaro, G. J.,** Complement-mediated lysis of type-C virus: effect of primate and human sera on various retroviruses, *Int. J. Cancer*, 21, 6, 1978.
107. **Oldstone, M. B. A. and Dixon, F. J.,** Acute viral infection: tissue injury mediated by anti-viral antibody through a complement effector system, *J. Immunol.*, 107, 1274, 1971.
108. **Welsh, R. M., Jr., Lampert, P. W., Burner, P. A., and Oldstone, M. B. A.,** Antibody-complement interactions with purified lymphocytic choriomeningitis virus, *Virology*, 73, 59, 1976.
109. **Welsh, R. M., Jr.,** Host cell modification of lymphocytic choriomeningitis virus and Newcastle disease virus altering viral inactivation by human complement, *J. Immunol.*, 118, 348, 1977.
110. **Geschwender, H. H., Rutter, G., and Lehmann-Grube, F.,** Lymphocytic choriomeningitis virus. II. Characterization of extractable complement-fixing activity, *Med. Microbiol. Immunol.*, 162, 119, 1976.
111. **deBracco, M. M. E., Rimoldi, M. T., Cossio, P. M., Rabinovich, A., Maiztegui, J. I., Carballal, G., and Arana, R. M.,** Argentine hemorrhagic fever. Alterations of the complement system and anti-Junin-virus humoral response, *N. Engl. J. Med.*, 299, 216, 1978.
112. **Suvatte, V., Pongpipat, D., Tuchinda, S., Ratanawongs, A., Tuchinda, P., and Bukkavesa, S.,** Studies on serum complement C3 and fibrin degradation products in Thai hemorrhagic fever, *J. Med. Assoc. Thailand*, 56, 24, 1973.
113. **Berry, D. M. and Almeida, J. D.,** The morphological and biological effects of various antisera on avian infectious bronchitis virus, *J. Gen. Virol.*, 3, 97, 1968.
114. **Gocke, D. J., Hsu, K., Morgan, C., Bombardieri, S., Lockshin, M., and Christian, C. L.,** Association between polyarteritis and Australia antigen, *Lancet*, 2, 1149, 1970.
115. **Gocke, D. J., Hsu, K., Morgan, C., Bombardieri, S., Lockshin, M., and Christian, C. L.,** Vasculitis in association with Australia antigen, *J. Exp. Med.*, 134, 330S, 1971.
116. **Kohler, P. F.,** Clinical immune complex disease. Manifestations in systemic lupus erythematosus and hepatitis B virus infection, *Medicine*, 52, 419, 1973.
117. **Alpert, E., Isselbacher, K. J., and Schur, P. H.,** The pathogenesis of arthritis associated with viral hepatitis. Complement component studies, *N. Engl. J. Med.*, 285, 185, 1971.
118. **Onion, D. K., Crumpacker, C. S., and Gilliland, B. C.,** Arthritis of hepatitis associated with Australia antigen, *Ann. Intern. Med.*, 75, 29, 1971.
119. **Wands, J. R., Mann, E., Alpert, E., and Isselbacher, K. J.,** The pathogenesis of arthritis associated with acute hepatitis-B surface antigen-positive hepatitis. Complement activation and characterization of circulating immune complexes, *J. Clin. Invest.*, 55, 930, 1975.
120. **Kosmidis, J. C. and Leader-Williams, L. K.,** Complement levels in acute infectious hepatitis and serum hepatitis, *Clin. Exp. Immunol.*, 11, 31, 1972.
121. **Charlesworth, J. A., Lawrence, S., Worsdall, P. A., Roy, L. P., and Boughton, C. R.,** Acute hepatitis. Significance of changes in complement components, *Clin. Exp. Immunol.*, 28, 496, 1977.
122. **Wands, J. R., Alpert, E., and Isselbacher, K. J.,** Arthritis associated with chronic active hepatitis: complement activation and characterization of circulating immune complexes, *Gastroenterology*, 69, 1286, 1975.
123. **Elias, E., Potter, B. J., Thomas, H. C., and Sherlock, S.,** C3 metabolism in HBsAg positive and negative chronic active liver disease (CALD), *Gut*, 17, 389, 1976.
124. **Eknoyan, G., Gyorkey, F., Dichoso, C., Martinez-Maldonado, M., Suki, W. N., and Gyorkey, P.,** Renal morphological and immunological changes associated with acute viral hepatitis, *Kidney Int.*, 1, 413, 1972.
125. **Knieser, M. R., Jenis, E. H., Lowenthal, D. T., Bancroft, W. H., Burns, W., and Shalhoub, R.,** Pathogenesis of renal disease associated with viral hepatitis, *Arch. Pathol.*, 97, 193, 1974.
126. **Kohler, P. F., Cronin, R. E., Hammond, W. S., Olin, D., and Carr, R. I.,** Chronic membranous glomerulonephritis caused by hepatitis B antigen-antibody immune complexes, *Ann. Intern. Med.*, 81, 448, 1974.

127. **Myers, B. D., Griffel, B., Naveh, D., Jankielowitz, T., and Klajman, A.,** Membrano-proliferative glomerulonephritis associated with persistent viral hepatitis, *Am. J. Clin. Pathol.*, 60, 222, 1973.
128. **Takekoshi, Y., Tanaka, M., Shida, N., Satake, Y., Saheki, Y., and Matsumoto, S.,** Strong association between membranous nephropathy and hepatitis-B surface antigenaemia in Japanese children, *Lancet*, 2, 1065, 1978.
129. **Takekoshi, Y., Tanaka, M., Miyakawa, Y., Yoshizawa, H., Takahashi, K., and Majumi, M.,** Free "small" and IgG-associated "large" hepatitis B e antigen in the serum of two patients with membranous glomerulonephritis, *N. Engl. J. Med.*, 300, 814, 1979.
130. **Cowan, K. M.,** Studies on the coliphage neutralizing activity of normal human serum, *J. Immunol.*, 88, 476, 1962.
131. **Toussaint, A. J. and Muschel, L. H.,** Studies on the bacteriophage neutralizing activity of serum. I. An assay procedure for normal antibody and complement, *J. Immunol.*, 89, 27, 1962.
132. **Muschel, L. H. and Toussaint, A. J.,** Studies on the bacteriophage neutralizing activity of serums. II. Comparison of normal and immune phage neutralizing antibodies, *J. Immunol.*, 89, 35, 1962.
133. **Adler, F. L., Walker, W. S., and Fishman, M.,** Amplification of phage neutralization by complement, antiglobulin, and anti-allotype sera, *Virology*, 46, 797, 1971.

Chapter 6

PARASITES

Primitive members of the animal kingdom, parasites may be classified as either Protozoa or Helminths (Metazoa). Discussion of those parasites or parasitic diseases for which there are data regarding the complement system will be organized around Table 1. The subject of complement activation by parasites has also been reviewed by Santoro et al.[1] Mahmoud[2] has emphasized immunologically critical differences between the Protozoa and Metazoa. Protozoa are more like bacteria, being unicellular and replicating within the host. Helminths on the other hand are multicellular and generally do not divide or multiply within the definitive host.

One characteristic of parasitic infections is chronicity. It is usual for parasitic infections to last months or years and, in many cases, to be associated with relatively little in the way of symptomatic disease; even when symptomatic disease develops, destruction of the host is rarely so rapid as in bacterial and viral diseases. In many cases a fairly satisfactory symbiotic relationship develops between host and parasite. Such a situation requires that there be some circumvention of the usual host immune responses. Ogilvie and Wilson[3] divided the mechanisms by which parasites evade the immune response into two: defects in host immune response and certain characteristics of the parasites that enable them to evade an otherwise effective host response. Host immune defects might include those on a genetic or developmental basis, such as an immature immune system in the very young. Acquired host immune defects may be induced by overwhelming numbers of parasites or by soluble immunoregulatory factors secreted by parasites. Parasitic antigens may become incorporated into immune complexes, which may themselves be immunosuppressive. Many parasites go through multiple developmental stages within the host; each transformation presents an antigenically new microbe with which the immune system must cope. Furthermore, some Protozoa spontaneously develop variant antigens in organisms derived from a single clone such that the host is confronted with an array of antigens, some of which can be cross-reactive but others of which are distinct to each variant. Among trypanosomes, antigenic change is such that totally new antigenic structures appear on the surface of the organisms approximately weekly. A final mechanism by which parasites evade a normal immune system is the surface accretion of host proteins, which camouflage parasitic antigens.

Several other immunological features characteristic of parasitic (especially Helminthic) infections include "allergic" symptomatology, induction of IgE antibodies, and eosinophilia.[4] The prominence of eosinophilia is especially intriguing, because eosinophils may have receptors for C3b as well as IgG.[5] The percentage of eosinophils expressing either or both of these receptors is normally lower than the percentage of polymorphonuclear neutrophils bearing such receptors.

I. PROTOZOA

Protozoa are primitive unicellular animals. Blood protozoa primarily produce intravascular infections, such as malaria, babesiosis, and trypanosomiasis; tissue protozoa are largely intracellular parasites and include the agents of toxoplasmosis and leishmaniasis; gut protozoa are usually confined to the gastrointestinal tract and include the agents of amebiasis and giardiasis. These organisms may produce their

Table 1
PARASITES AND PARASITIC DISEASES OF MAN

Protozoa
  Plasmodia (malaria)
    P. falciparum, P. vivax, P. ovale, P. malariae
  Babesia
    B. microti (babesiosis)
  Trypanosoma
    T. rhodesiense (East African trypanosomiasis)
    T. gambiense (West African trypanosomiasis)
    T. cruzi (Chagas' disease)
  Pneumocystis carinii
  Leishmania
    L. tropica (Oriental sore)
    L. braziliensis (New World cutaneous leishmaniasis, espundia)
    L. mexicana (New World cutaneous leishmaniasis, Chiclero ulcer)
    L. donovani (visceral leishmaniasis, kala-azar)
  Entamoeba histolytica (amebiasis)
  Giardia lamblia (giardiasis)
  Trichomonas vaginalis (trichomoniasis)
  Balantidium coli (balantidiasis)
  Isospora belli (coccidiosis)
  Naegleria species (primary amebic meningoencephalitis)
  Toxoplasma gondii (toxoplasmosis)
Helminths
  Cestodes (tapeworms)
    Diphyllobothrium latum (fish tapeworm)
    Taenia saginata (beef tapeworm)
    Taenia solium (pork tapeworm, cysticercosis)
    Hymenolepis nana (dwarf tapeworm)
    Hymenolepis diminuta (rat tapeworm)
    Dipylidium caninum (common tapeworm of cats and dogs)
    Echinococcus granulosus, E. multilocularis (hydatid disease)
  Trematodes (flukes)
    Schistosoma mansoni, S. japonicum, S. haematobium (schistosomiasis)
    Liver flukes
      Clonorchis sinensis
      Opisthorchis viverrini, O. felineum
      Dicrocoelium dendriticum
      Fasciola hepatica
    Lung flukes
      Paragonimus westermani (paragonimiasis)
    Intestinal flukes
      Fasciolopsis buski
  Nematodes (roundworms)
    Intestinal
      Strongyloides stercoralis (strongyloidiasis)
      Capillaria species (capillariasis)
      Necator americanus (hookworm disease)
      Ancylostoma duodenale (hookworm disease)
      Ancylostoma brasiliensis (cutaneous larva migrans)
      Trichostrongylus species
      Gnathostoma species
      Oesophagostoma species
      Ascaris lumbricoides (ascariasis)
      Toxocara catis, T. canis (visceral larva migrans)
      Anisakis marina
      Trichuris trichiura (trichuriasis, whipworm infection)
      Enterobius vermicularis (enterobiasis, pinworm infection)

Table 1 (Continued)

Tissue
   Trichinella spiralis (trichinosis)
   Angiostrongylus cantonensis
   Filaria species (filariasis)

disease by direct tissue destruction; on the other hand, the disease may be largely a reflection of the immune response of the host.[2]

A. Plasmodia (Malaria)

Plasmodial sporozoites are transferred by the mosquito from the infected blood of one host to that of another. From the blood stream sporozoites quickly infect liver cells. After a period of development, organisms reenter the blood stream and begin the characteristic cycles of erythrocyte parasitism and intermittent hemolysis. These blood stages of the parasite are those most likely to affect and be affected by the immune system of the host.[6,7] Some of the forms of malaria also have a secondary exoerythrocytic (hepatic) stage, characterized by dormancy and isolation from the host defense system.[6] About a week following the initial parasitemia, an antibody response is first demonstrable and is associated with onset of clinical disease and hematological changes. Parasitemia declines following antibody production, but infection may not be eradicated if organisms are sequestered in the liver. However, antibodies are protective.[7] Cohen and Butcher[8,9] demonstrated that immune serum could kill or inhibit *Plasmodium knowlesi* organisms released from ruptured, infected erythrocytes, but had no effect on those organisms already within erythrocytes. The major effect of antibodies was in preventing reinfection of erythrocytes during succeeding cycles of erythrocyte rupture and parasite release. In their studies the antibodies (IgG and IgM) were complement independent. Likewise, in experiments in rats with *P. berghei*, Diggs et al.[10] could show no effect of cobra venom factor treatment on parasitemia or on the protective role of passively transferred serum. Furthermore, among inbred strains of mice, C5 deficiency did not increase susceptibility to *P. berghei* infection.[11]

However, complement-fixing antibodies appear soon after initial parasitemia[6] and complement appears to be involved, though perhaps not protectively, in various stages of malaria. Roy and Mukerjee[12] described "sharp" decreases in hemolytic complement during terminal stages of disease in monkeys infected with *P. knowlesi*. Dulaney[13] measured hemolytic complement activity in the sera of 27 patients with neurosyphilis before and after malaria (induced therapeutically to produce fever). He noted a decrease in complement that ranged from very slight to very severe but parallelled the severity of the induced malaria. Fogel, Cooper, and colleagues[14,15] produced experimental malaria in rhesus monkeys (*P. knowlesi*), hamsters (*P. berghei*), and chickens (*P. gallinaceum*). They demonstrated decreases in hemolytic complement activity that were directly correlated with the percentage of circulating erythrocytes containing parasites. Furthermore, declines in complement levels occurred daily following release of parasites from the erythrocytes into the blood. In the monkeys infected with *P. knowlesi*, the components C1, C2, and C3 were all profoundly depressed, again proportionally to the degree of parasitemia.[15] These changes in complement levels appeared within four days of infection; the authors questioned whether such an early response could be entirely related to the appearance of antibody.[15]

Topley et al.[16] demonstrated increased immunoconglutinin levels in six of ten patients with malaria (immunoconglutinin is an antibody directed against neoantigens on the C3b molecule). Children with *P. falciparum* infections studied by Greenwood and Brueton[17] had marked depressions of C4 and C1q with less pronounced depression of C3; factor B was modestly if at all depressed. Component levels were lower in those children with severe degrees of parasitemia. Srichaikul et al.[18] demonstrated not only a decreased concentration of C3 but also a more rapid clearance of radiolabeled C1q in patient with *P. falciparum* malaria. Again values were lower in patients with parasitemia. Similar depression of C3 was noted in children with *P. falciparum* malaria by Ree.[19] Petchclai et al.[20] detected complement activation in less than one third of patients with acute falciparum malaria; lowest levels were associated with jaundice, which may be a correlate of parasitemia. Rose et al.[21] also documented increased levels of immunoconglutinin in cattle infected with *Anaplasma marginale*, a bovine plasmodium.

In patients and volunteers with *P. vivax* infection, Neva et al.[22] found the declines in hemolytic complement activity (CH50) and C4 were coincident with the clinical paroxysm characterized by fever and headache and associated with rupture of the parasite-laden erythrocytes and subsequent parasitemia. The falls in C4 and CH50 correlated with the presence of complement-fixing antibody and with the degree of parasitemia. They noted minimal changes in C3 and C6. Subsequent studies[23,24] of rhesus monkeys infected with *P. coatneyi* confirmed that changes in complement levels were dependent on production of antibody and very clearly followed erythrocyte rupture. Activation appeared to involve almost exclusively the early components of the complement system.

Thus complement activation, while apparently not protective, is demonstrable and may play some role in the immunopathology of malaria. Several postulated mechanisms for the prominent anemia could involve complement. Connal[25] in 1912 demonstrated phagocytosis of parasitized erythrocytes in malaria. He was not able to establish that an opsonin was involved. Plasmodial antigens appear on the surface of infected as well as well as noninfected erythrocytes phagocytized in liver, spleen, and bone marrow; in vitro erythrophagocytosis occurs in the presence of antiserum to these antigens.[26] Furthermore, infected erythrocytes can be agglutinated and lysed by antiserum to the plasmodial antigens.[26] A role for complement in these observations is speculative. Glew et al.[23] suggested that complement might play an active role in rupture of erythrocytes rather than simply being activated secondary to that rupture. Topley et al.[16] demonstrated C3b on the surface of erythrocytes of patients with malaria. Thus complement activation may be related in a causal way to the erythrocyte rupture and the extravascular phagocytosis of parasitized and perhaps nonparasitized erythrocytes.

Disseminated intravascular coagulation is a recognized complication of malaria, especially falciparum malaria. Greenwood and Brueton[17] demonstrated an inverse correlation between levels of complement and fibrin degradation products (a measure of intravascular coagulation). Likewise Srichaikul et al.[18] had correlated depressions of C3 with thrombocytopenia and disseminated intravascular coagulation. Neva et al.[7] have however suggested that the thrombocytopenia is due not to disseminated intravascular coagulation but rather to platelet sequestration in the spleen, as a result of opsonization by antibody and perhaps complement.

Nephritis and the nephrotic syndrome have long been recognized as prominent features of chronic malaria, primarily chronic infection with *P. malariae*.[6] In rhesus monkeys splenectomized and infected with *P. cynomolgi*, plasmodial antigens as well as "faint" deposits of immunoglobulin and C3 were demonstrated in renal endothelium.[27] Soothill and Hendrickse[28] noted the epidemiological association of childhood nephrotic syndrome with *P. malariae* infection in Nigeria. They[28] demonstrated immune complexes, containing C3, in serum from such patients. In immunopathological studies of renal biopsies of such patients,[29] immunoglobulin and complement were demonstrable in all kidneys. Elution studies in one patient suggested that the antibody was specific for *P. malariae*. These authors[29] postulated that the propensity of *P. malariae* infection to lead to nephrosis was due to the fact that this disease is characterized by a chronic and "prolonged release of antigen of exoerythrocyte source," with minimal antigenic variation. Such a disease, they suggested, would fulfill Dixon's criteria for the type of antigenic exposure perfectly suited to induce immune-complex nephritis. Ward and Kibukamusoke[30] obtained similar results in approximately one third of patients with *P. malariae* infection. Monkeys experimentally infected with *P. cynomolgi* also developed granular renal deposits of immunoglobulin and complement within nine days of infection.[31] In studies with experimental *P. berghei* infections in mice, June et al.[32] demonstrated similar nephritis and circulating immune complexes. Immune-complex nephritis has also been seen with acute falciparum malaria;[33] however, changes in serum complement activity were not demonstrated.

B. Babesia (Babesiosis)

Babesiosis is similar in many respects to malaria. Studies of the complement system and infection with *Babesia microti*, the etiologic agent of human babesiosis, were not identified in this review. Studies have been done however with a rat parasite, *B. rodhaini*, primarily because of the resemblance to malaria. *B. rodhaini*, however, does not undergo sexual reproduction in the host, and thus changes noted are those of the *initial* infection.[34]

Babesia infections in rodents are associated with hypocomplementemia[35,36] and increased catabolism of C3.[35] Such changes could not be correlated with increased catabolic rates of immunoglobulins.[35] Decreases in complement components generally correlated with levels of parasitemia.[36] Hemolytic function of the alternative pathway was not affected, suggesting that activation was primarily through the classical pathway.[36] This complement activation appeared to correlate with a glomerulonephritis mediated by immune complexes and complement.[35,37]

Chapman and Ward[34] suggested a novel function of complement in the pathophysiology of babesiosis. They incubated human erythrocytes, *B. rodhaini*, and variously treated human sera. After incubation they measured the degree of parasitemia (i.e., the percentage of erythrocytes infected in vitro). In the absence of serum, no erythrocytes were infected. Human serum supported approximately a 10% parasitemia. EGTA did not inhibit that effect of serum; whereas EDTA chelation, zymosan or cobra factor pretreatment, heat inactivation, or other treatments that decreased the concentration of C3 all inhibited the effect of normal serum. Neither was C3-deficient serum supportive of erythrocyte invasion by parasites, but activity could be restored to that serum by exogenous C3. Properdin seemed important, as did some of the terminal components, such as C5. The authors postulated that babesia activated the alternative pathway with deposition of C3 on their surface. C3 could then mediate attachment of the organisms to erythrocytes; subsequent internalization may have been partially mediated by erythrocyte membrane damage from terminal complement components.

C. Trypanosomes

The trypanosomes are arthopod-borne parasites characterized morphologically by flagella. The parasites exist extracellularly in blood, lymph, or cerebral spinal fluid. In addition, *Trypanosoma cruzi* may also exist in a tissue phase intracellularly.[38-40]

1. African Trypanosomiasis

Three different species of trypanosomes, very closely related and transmitted by the tsetse fly, appear to cause this syndrome. *T. brucei* infects mammals but rarely man. Two pathogens are morphologically indistinguishable from *T. brucei* and are often considered part of the "brucei-complex:" *T. gambiense* (the etiologic agent of West African trypanosomiasis) and *T. rhodesiense* (the etiological agent) of East African trypanosomiasis).[39,40] West African trypanosomiasis is a chronic disease characterized by fever and gradually progressive meningoencephalitis, leading to eventual stupor and coma. East African trypanosomiasis is a much more acute and rapidly progressive meningoencephalitis leading to death within a few weeks.[40]

Complement involvement in the syndromes of African trypanosomiasis was suggested by studies in animals infected with *T. brucei*.[41] Development within 30 days of infection of antibodies to immunologically altered C3 (i.e., C3b) suggested activation of C3. Barrett-Connor et al.[42] studied a patient who developed African trypanosomiasis complicated by disseminated intravascular coagulation. During the first week of illness, the hemolytic complement level was 10 units per mℓ (normal, 40 to 60 units per mℓ); C3 concentration was 94 mg/dℓ (normal, 100 to 190 mg/dℓ). Greenwood and Whittle[43] reported on 60 patients with West African sleeping sickness, in various stages of chronic disease. In general, levels of C3, C4, and factor B were all depressed and some patients had circulating C3 conversion products. No patients however had clinical features of immune-complex disease. Assoku et al.[44] postulated that the complement activation in trypanosomiasis, by both the classical and alternative pathways, led to a functional C3 deficiency with impaired immune responses and impaired antigen processing by the reticuloendothelial system.

Musoke and Barbet[45] identified surface components of *T. brucei* that were able to activate the complement system. Particulate fractions of the organism could activate complement, in serum chelated with EGTA, through the alternative pathway. A glycoprotein from the surface, approximately 64,000 d weight, was capable of decomplementing serum in vitro in a dose-dependent fashion. The early complement components, C1, C4, and C2, were preferentially consumed and complement consumption was blocked by EGTA. These glycoprotein surface antigens, which may be shed into the host blood and tissues, undergo repeated variation, with replacement by antigenically dissimilar glycoproteins.[46]

Anemia is a characteristic feature of African trypanosomiasis.[47] Hemolysis mediated by antibody and complement appears to play a role in this anemia. Woodruff et al.[48] demonstrated shortened erythrocyte survival time and sequestration of labeled erythrocytes in the spleen. Furthermore, erythrocytes from patients with trypanosomiasis had surface C3; two patients tested had increased antibodies to immunologically activated C3 (immunoconglutinin). Tizard et al.[46] suggested that glycoprotein surface antigens of trypanosomes bound directly to erythrocytes might activate complement and mediate the anemia.

Mice infected with *T. brucei*[47] and monkeys infected with *T. rhodesiense*[49] developed an immune-complex nephritis associated with glomerular deposition of C3 and immunoglobulin. Nagle et al.[49] correlated these renal changes with hypocomplementemia.

Observations have been made regarding the potential protective role of complement in trypanosomiasis. Yorke et al.[50] observed in 1930 that normal human serum was trypanocidal in vitro against a variety of strains of trypanosomes, including a laboratory strain of *T. rhodesiense* and several strains pathogenic only for animals other than man. A laboratory strain of *T. gambiense* or a recently isolated strain of *T. rhodesiense* was not subject to this effect of serum. The factor(s) was heat labile but only at 60°C (not at 56°C). Corson[51] further analyzed the serum susceptibility of a strain of *T. rhodesiense*. The organism freshly isolated from man was serum resistant but became serum sensitive following prolonged passage in lower animals. The serum-sensitive strain was less virulent but could still infect the author. Adams[52] further demonstrated that strains of *T. gambiense* and *rhodesiense* pathogenic for man were invariably resistant to human serum, even though they might be sensitive to serum from other animals such as sheep. *T. brucei* strains, on the other hand, were sensitive to the lytic effect of human serum. Aaronovitch and Terry[53] attributed human resistance to infection to this lytic activity. They furthermore presented evidence that the trypanolytic factor was related primarily to IgM antibody. Diggs et al.,[54] however, suggested that antibody alone could not impair metabolic activity of trypanosomes (*T. rhodesiense*) but required also a heat-labile factor present in fresh nonimmune serum.

Balber et al.[55] demonstrated an interesting interaction of trypanosomes (*T. brucei* and *T. congolense*) with immunoglobulins and complement. Incubation of the organisms with specific antibody at 0°C followed by addition of guinea pig complement at 37°C resulted in lysis. However, if the antibody and the organisms were incubated at 37°C for 30 min prior to the addition of complement, no lysis ensued. They suggested that the 37°C incubation resulted in elimination of the potentially lytic surface antigen-antibody complexes by a mechanism related to the phenomenon of antigenic variation; i.e., antigen-bound antibody was shed when the antigen was replaced by a new variant antigen. When fresh antibody was added after the shedding of the previous antibody, the organisms could again by lysed.

Flemmings and Diggs[56] found that strains of *T. rhodesiense* may be damaged by leukocytes in a fashion dependent on antibody and complement and reminiscent of antibody-dependent cellular cytotoxicity demonstrated in other systems (see Chapter 1). C4-deficient guinea pig serum was as effective as normal guinea pig serum and the activity could not be abolished by EGTA, which data indicated that the cytotoxicity was mediated through the alternative pathway.

2. Experimental Infections with Nonhuman Trypanosomes

T. congolense produces a trypanosomiasis of cattle. Kobayashi and Tizard[57] demonstrated that C3 levels in experimentally infected calves declined from a preinfection value of 3.75 mg/mℓ to a low value of 0.95 mg/mℓ at 2 weeks after infection. Values remained low through the 18th week of infection. The lowest values coincided with peak levels of complement-fixing antibody. Nielsen and colleagues[58,59] noted not only decreases in C3 but also decreases in total hemolytic complement activity, C1 and C1q. Since these changes were associated with increased catabolic rates for C1 and C3, the lowered complement values appeared to reflect consumption rather than decreased synthesis. Properdin levels fluctuated with cycles of disease but were not clearly abnormal at any time, and C8 values were consistently normal.

As with African trypanosomiasis in humans, infection with *T. congolense* produces an anemia in cattle. The anemia is hemolytic and characterized by a positive Coombs test. Unlike the case in human diseases, erythrocytes from infected cattle only occasionally had surface C3.[60]

T. congolense directly activated bovine complement, apparently through the classical pathway by direct antibody-independent activation of C1.[61,62] *T. congolense* could not activate guinea pig or human complement.[61] A pathogen of rats, *T. lewisi*, similarly activated human, guinea pig, and bovine complement.[61,62] The complement activating activity was attributable to at least two trypanosomal products, which were suggested to represent means of diverting the complement system.[63] Greater parasitemia was seen in rats depleted of complement not only by cobra venom factor but also by administration of factors produced in culture by *T. lewisi*.[64]

Lange and Lysenko[65] demonstrated phagocytosis of *T. lewisi* by rat peritoneal exudate cells. Immune serum enhanced phagocytosis whereas nonimmune serum did not. The factor in immune serum was absorbable and heat labile (46°C, 30 min). Jarvinen and Dalmasso[66] explored the antibody response to experimental *T. lewisi* infection in rats. They identified, early in the course of infection, a complement-independent antibody that inhibited reproduction and converted reproducing trypanosomes to "monomorphic adults." The second antibody, produced later in infection, was trypanocidal through complement-dependent lytic and opsonic mechanisms; its appearance was associated with lowered serum complement levels (CH50, C4, C3). The lowest complement values correlated with maximum parasitemia. C4-deficient animals were not different from normals in terms of degree of parasitemia, percentage of parasites in reproductive form, duration of infection, rate of parasite elimination, or changes in C3 levels. In contrast to other studies,[64] these authors[66] found cobra venom factor treatment not to affect any variables in normal or C4-deficient rats. These authors[66] concluded that, while complement might be activated intensely during infection, there was little evidence to suggest any important role for complement in recovery.

Dusanic[67] and Jarvinen and Dalmasso[68] reached similar conclusions using *T. musculi* infections in mice. C5-deficient mice did not differ from normals in parasitemia, reproductive activity of the trypanosomes, or clearance of organisms. These data suggested that complement-mediated cytolysis was not required for normal in vivo defenses. C3 depletion with cobra venom factor slightly reduced the rate of elimination of adult parasites and prolonged infections but differences were not great.[68] However, in other studies, Jarvinen and Dalmasso[69] demonstrated that the anemia of *T. musculi* infection in mice could be correlated with the presence of immunoglobulin and complement on the surface of erythrocytes. The fact that C5-deficient mice were similar in degree of anemia and of erythrocyte-bound C3 suggested that intravascular hemolysis played little role and that the anemia was most likely due to erythrophagocytosis by the reticuloendothelial system.[69]

Taliaferro[70] studied trypanolytic antibodies developing during experimental infection of rats with a similar trypanosome, *T. duttoni*. This trypanolytic activity was heat labile. Activity could be restored to immune serum by fresh serum.

In vitro studies[71] of the interaction of a simian pathogen, *T. cyclops*, with normal human serum indicated that the organisms could be lysed by factors that were heat labile at 56°C or 52°C and inhibited by EDTA chelation or cobra venom factor, but not inhibited by C2 deficiency or MgEGTA chelation. Absorption with the organism did not inhibit lytic activity. The authors concluded that the trypanosomes were lysed by the alternative complement pathway.

3. *Trypanosoma cruzi* (Chagas' Disease)

Chagas' disease is endemic through most of South and Central America and the southern portions of North America. The arthropod vector is the reduviid bug. Acute infection may be asymptomatic or associated with periorbital swelling and fever. Trypanosomes subsequently infect predominately the cells of the smooth muscle of the heart or gut.[40,72]

Mice infected with *T. cruzi* developed hypocomplementemia by day 30 of infection; complement values remained low throughout subsequent infection.[73] Complement activation in vitro in human and guinea pig sera appeared to be due to a soluble factor that apparently activated, at least partially, the alternative pathway.[73] Furthermore, mice treated with cobra venom factor following experimental infection demonstrated more rapid death, greater mortality, and a greater parasitemia.[74]

In vitro studies on the mechanism of any protective role for the complement system in defense against *T. cruzi* have been conflicting. In 1943, Denison[75] observed that rats recovered from trypanosomiasis were immune. Fresh serum from such animals lysed *T. cruzi* in vitro. This lytic effect was dependent on a heat-labile factor. However, heat-inactivated serum could still immobolize, though not lyse, the trypanosomes. Cultural forms of the trypanosome (epimastigotes) can be lysed by normal human and guinea pig serum.[76,77] There is controversy whether this lytic activity requires the classical[76] or the alternative[77] pathway. Virulent or infective forms of the trypanosomes are generally thought to be resistant to lysis by human serum; on the other hand, sera of animals innately resistant to infection with *T. cruzi*, i.e., birds and frogs, are capable of lysing these organisms.[74] However, Budzko et al.[74] were able to demonstrate in vitro lysis of virulent forms of *T. cruzi* by human serum. Lysis required antibody and complement. EGTA partially inhibited lysis, indicating some requirement for the classical pathway.

Kierszenbaum and colleagues[78] studied mechanisms of resistance of animals innately resistant to *T. cruzi* infection. Parasites were maintained in mice and then studied with chicken serum. Chicken serum lysed these virulent *T. cruzi* in vitro. The lytic activity was heat labile and inhibited by cobra venom factor treatment and EDTA chelation, but not by EGTA chelation. Chickens depleted of complement by cobra venom factor had diminished resistance to infection; trypanosomes circulated for prolonged periods of time (greater than 24 hr in complement-depleted animals compared to less than 4 min in complement-sufficient animals). Parasitemia in mice could be diminished by transfusion of normal chicken serum. The normal lytic activity of chicken serum did not appear to be dependent on immunoglobulin depletion as accomplished by neonatal bursectomy, cyclophosphamide treatment, or neonatal infection with "infectious bursal disease virus." In contrast, in mice, Kierszenbaum[79] demonstrated no lysis by normal mouse serum but lysis by immune mouse serum. The trypanolytic effect of immune mouse serum was dependent on complement, apparently the alternative pathway, and a cross-reactive antibody. Appropriately, complement depletion at the time of antibody production (9 days after infection) led to exacerbation of disease, with increased parasitemia and earlier mortality. These results suggested that animals innately resistant to *T. cruzi* infection were capable of lysing the infective, virulent forms of the organism without the participation of antibody. In contrast, susceptible animals required an antibody in order to mediate this lytic effect. The fact that the antibody was cross-reactive among strains of trypanosomes may offer some explanation for the variability among experimental observations.

D. Leishmania

Leishmania are intracellular parasites transmitted from man to man by sandflies. They produce cutaneous (*Leishmania tropica*), mucocutaneous (*L. braziliensis* and *L. mexicana*), or visceral (*L. donovani*) infections. Visceral leishmaniasis is also known as kala-azar.[80,81]

The cutaneous and mucocutaneous infections by *L. tropica, L. braziliensis,* and *L. mexicana* generally elicit primarily cellular immune responses. Visceral leishmaniasis is characterized by diffuse parasitism and a syndrome of fever, hepatosplenomegaly, anemia, leukopenia, hyperglobulinemia, and chronic debilitation. During active infection, there is no demonstrable delayed hypersensitivity.[82] Only a few studies have evaluated the role of complement in these infections.

In ten patients with kala-azar, Rezai et al.[83] demonstrated, among many other variables, depressions of serum complement in two patients, both of whom had proteinuria. Two other patients had borderline low C3 values. And Woodruff et al.[84] attributed the anemia of visceral leishmaniasis to complement and perhaps antibody on erythrocytes and resultant splenic sequestration. They demonstrated elevated levels of antibodies to C3b as well as C3 on erythrocyte surfaces.

Adler[85] showed that normal serum could rapidly lyse *L. donovani* in vitro. Similar findings have been demonstrated for other species of leishmania. Lytic activity was also present in rabbits, guinea pigs, and rats. The factors were both heat labile and absorbable and were thought to be antibody and complement.[86-88] However, experimental subjects developed leishmaniasis in spite of readily demonstrable serum bactericidal activity.[85] Mauel and Behin[82] suggested that the organisms may be able to penetrate host cells so rapidly as to avoid lytic effects of serum.

E. *Entamoeba histolytica*

Entamoeba histolytica is a pathogen primarily of the lower gastrointestinal tract and causes amebic dysentery. Occasionally the organism will produce a liver abscess or some other form of invasive disease. Forty-three patients with amebic abscesses and 13 patients with amebic colitis studied by Ganguly et al.[89] had normal or elevated complement levels (CH50 and C3).

Ortiz-Ortiz et al.[90] studied in vitro interactions of *E. histolytica* and the complement system. They demonstrated that the organisms could activate complement, consuming primarily C3 with minimal if any consumption of the early components of the classical pathway. Upon incubation in normal human serum, viable organisms became less motile after 5 min. They subsequently became rounded and developed cytoplasmic vacuoles; after 20 to 30 min, lysis occurred. This cytopathic effect of normal serum was subject to dilution and appeared to be mediated through the alternative pathway of complement; activity was heat labile, required magnesium but not calcium, and was destroyed by cobra venom factor or zymosan treatment. Immune serum showed enhanced activity that could be absorbed by organisms; however, activity of normal serum could not be absorbed by the amebae.

F. *Toxoplasma gondii*

Toxoplasma gondii is an intracellular parasite that causes frequent infection but uncommon disease. Organisms are widely disseminated in the body but characteristically produce disease primarily in the central nervous system, lymph nodes, or the reticuloendothelial system. The disease is becoming more important as a pathogen of immunocompromised patients.[91] C5-deficient mice challenged intraperitoneally with *T. gondii* actually had a slightly better mortality rate than did normal mice, but factors other than complement seemed to be important in the pathogenesis of this disease.[92]

In 1948, Sabin and Feldman[93] observed that viable toxoplasma could be stained by a dye (methylene blue) whereas toxoplasma lysed by immune serum would not stain with the dye. This observation was of course the basis for the Sabin-Feldman dye test used in measuring antibody to *T. gondii*. In their studies, the effect of the immune serum was heat labile. They suggested that two factors were involved: antibody and an "accessory factor" which might be complement. Subsequent studies by Gronroos[94] and Feldman[95] indicated that the accessory factor might be the al-

Table 2
SITES OF INFECTION OF SOME HELMINTHS[a]

Intestinal
    Taenia saginata
    Fasciolopsis buski
    Trichuris trichiura
    Enterobius vermicularis
Tissue-intestinal
    Fasciola hepatica
    Strongyloides stercoralis
    Necator americanus
    Ancylostoma duodenale
    Ascaris lumbricoides
Tissue
    Taenia solium
    Echinococcus granulosus and *multilocularis*
    Schistosoma mansoni, japonicum, haematobium
    Toxocara canis
    Trichinella spiralis
    Filaria species

[a] Adapted from Warren.[4]

ternative complement pathway. Feldman[95] depleted serum of properdin by incubation at 17°C with zymosan and demonstrated that such treatment abrogated the lytic activity of immune serum. He could restore activity by addition of purified human or bovine properdin. In his studies properdin alone was not lytic, although Gronroos[94] suggested that properdin alone could damage toxoplasma.

II. HELMINTHS (METAZOA)

A characteristic of Helminthic infections is that the parasites, once in the body, do not multiply. Each egg or larva can become only one adult. Thus, as Warren[4] has pointed out, host immunological reactions may be helpful in limiting the number of larvae that can become adults, but "A severe host reaction to an organism that does *not* multiply must be largely harmful . . ." to the host. Warren[4] has divided Helminthic parasites into those that exist almost entirely in the gut and produce little immunological reaction, those that have both tissue and intestinal phases, and those that exist largely in tissue (though some gut stages may be involved) (Table 2). Helminth evasion of host immune systems has been discussed by Ogilvie and Wilson;[3] and interactions of Helminths with the complement system have been reviewed by Santoro et al.[1]

Lumsden[96] has reviewed the structure of these complex multicellular organisms. Like all cells, the outer boundary of these organisms is a cell membrane composed of a typical lipid bilayer with proteins and small amounts of carbohydrates. External to but an integral part of this cell membrane is the glycocalyx, composed of glycoproteins or glycolipids and oligosaccharides or polysaccharides. Some Helminths may have an acellular protein exoskeleton called a cuticle. There is contradictory data as to whether the cuticle is truly exterior to the plasma membrane or is actually intracellular, but Lumsden[96] thought it to be an integral cellular component. The surface coat of Helminths may mask surface antigens or may adsorb host antigens.

A. Cestodes (Tapeworms)

Cestodes have two life stages. Adult tapeworms live in the gastrointestinal tract and produce local symptoms. In some hosts, larvae may gain access to tissues where they develop into cysts. Symptoms develop primarily from local growth of these cysts.[97]

1. Taenia

Humans acquire taeniasis by ingestion of undercooked beef (*Taenia saginata*) or pork (*T. solium*). Viable cysts develop in the gastrointestinal tract into mature adults, from which egg-laden proglottids detach and are excreted. Following ingestion of eggs in human feces by other animals, larvae penetrate the gastrointestinal mucosa, migrate to tissue, usually liver or muscle, and develop into cysts. In humans, only *T. solium* may have this extra-intestinal phase.[97]

In experimental animals infected with *T. taeniaeformis*, two different antibody responses were demonstrated.[98] The first antibodies formed were directed against the infective larvae, i.e., larvae developed from the egg and capable of penetrating the gastrointestinal mucosa. Later, antibody developed against encysted larvae and was no longer reactive with the invasive form. Both sera induced increased larval permeability with leakage of intracellular molecules; complement was required for the reaction. Similar antibodies were induced by immunization of rabbits with *T. pisiformis*.[99] Musoke and Williams[100] demonstrated that these antibodies were protective in vivo by passive transfer of immunity to *T. taeniaeformis* in rats. Passive immunity was complement dependent as evidenced by abrogation of protective effect in cobra venom factor-treated rats. Mitchell et al.[101] obtained similar results in mice and noted also that mouse strains deficient in C4 or C5 were generally more susceptible to infection with *T. taeniaeformis*. When Hustead and Williams[102] measured permeability changes in immunologically damaged larvae, they found that leakage of ribonuclease A was increased by immune serum (complement dependent), whereas permeability to bovine serum albumin was not increased. Enhanced permeability of only the smaller molecule suggested that lesions produced in the cestodes might be analogous to more standard complement lesions on erythrocytes.

After about 5 days of growth in vivo, larvae are no longer sensitive to damage by antiserum.[100] Such a resistance may be related to development of complement inactivators by the larvae.[102] Furthermore, *T. taeniaeformis* cyst fluid activated complement in vitro and in vivo.[103] Such anticomplementary activity of cyst fluid was present as early as 20 days after infection[108] and was associated with polyanionic macromolecules.[104] Nevertheless, complement activation during experimental infection could not be demonstrated.[103]

2. Hymenolepis

Hymenolepis nana is infective for man. Adult forms and larvae live in the intestinal tract, eggs secreted by the adult worms developing into larvae within the intestinal tract and completing the cycle.[97] Heyneman and Welsh[105] demonstrated profound toxic effects of rabbit antiserum against *H. nana* in vitro. Though they did not investigate the effect of complement, their antiserum was prepared so as to preserve complement. Befus and Threadgold[106,107] demonstrated C3 antigens on the tegument of *H. diminuta* during in vivo infection in mice.

3. Echinococcus

The tapeworms *Echinococcus granulosus* and *E. multilocularis* exist as adults in the intestines of carnivores such as dogs, foxes, and wolves. Eggs excreted in the feces may contaminate human food or be otherwise accidentally ingested. Within the human intestinal tract, eggs develop into infective larval forms that penetrate the mucosa. By way of mesenteric blood vessels, larvae are carried throughout the body but especially to the liver, where they develop into cysts. As cysts gradually increase in size, new organisms (called scolices or protoscolices) form in large numbers from the germinal lining. The cyst is called a hydatid cyst.[97]

Protoscolices have been shown to be lysed by normal or immune serum from a variety of animal species including humans.[108-110] Lysis was inhibited by heat inactivation, cobra venom factor treatment, or EDTA chelation, these results indicating a requirement for complement.[109] Data as to the pathway activated have been conflicting. In some studies, absorption at 0°C with protoscolices did not inhibit lysis, though these studies were done with high concentrations (50%) of serum.[108] Likewise, Rickard et al.[110] demonstrated lysis in C4-deficient guinea pig serum. These studies would indicate independence of immunoglobulin or the classical pathway. On the other hand, Kassis and Tanner[111] inhibited lysis with either EDTA or EGTA; these data would indicate a requirement for the classical pathway. Explanation for the discrepancy among these data is unclear; but the two studies indicating alternative pathway function[108,110] were done with *E. granulosus*, whereas the study indicating classical pathway activation[111] was done with *E. multilocularis*. When studied by electron microscopy, the effect of serum on protoscolices was characterized first by bubbling of the tegument, followed by disintegration of the plasma membrane and subsequent autolysis.[112]

Host proteins, including immunoglobulin and complement, accumulate on the outer membrane of the cyst.[111] Many of these proteins, including complement proteins, are able to diffuse into the cyst fluid. Kassis and Tanner[109] showed that nonlarval contents of the hydatid fluid were able to activate complement. They suggested that this complement activation might be important in protecting the larvae.

Kassis and Tanner[113] also performed studies in vivo with *E. multilocularis* in rats. Normal rats infected intraperitoneally with protoscolices developed a mean cyst weight in the liver of 0.1 mg. Rats pretreated with cobra venom factor had an increased mean cyst weight of 10 to 14 mg. These authors suggested that the number of organisms that would reach the liver reflected complement function and the role of complement in defense against these parasites was in preventing larvae from surviving the transit from the gut to the liver.

B. Trematodes (Flukes)

Trematodes live either in the tissue of the host or in the gastrointestinal tract; larvae may also have a distinct tissue phase. The only studies dealing with complement and trematodes relate to schistosomiasis. Phillips and Colley[114] have reviewed host defenses and immunopathology of this disease. There are three species of schistosomes pathogenic for man; *Schistosoma mansoni*, *S. japonicum*, and *S. haematobium*. Adults of *S. mansoni* and *S. japonicum* live in mesenteric veins; adult *S. haematobium* worms live in the urinary bladder venous plexus. In either location, the worms mate and produce eggs. The majority of eggs pass through the intestinal or bladder wall and into the excreta. In fresh water, eggs develop into free-swimming larvae that infect snails, the intermediate host. After multiple developmental stages, the larvae leave the snail as cercariae that can penetrate intact human skin. The skin penetration involves a complex change in structure from cercaria to schistosomulum. The schistosomula migrate in the venous system through the lungs and return via the pulmonary veins to the liver, where they mature. Adults then migrate into mesenteric veins, mate, and begin egg laying. Some of the eggs produced in the mesenteric veins may embolize to the lung and liver rather than passing through the intestinal wall. These metastatic eggs elicit granulomatous reaction and fibrosis. Clinical symptoms may occur at various stages in this life cycle. At the site of penetration of cerceriae through the skin, a local dermatitis may develop, especially in patients previously infected. Heavy infections may result in pulmonary symptoms during the initial schistosomula passage through the lungs. Some heavy infections may, at the time of primary egg laying, result in a generalized syndrome of gastrointestinal disturbances, diffuse rash, hepatosplenomegaly, eosinophilia, and fever. The chronic phase of the disease is associated with progressive liver failure and portal hypertension due to hepatic fibrosis, basically resulting in a picture of chronic cirrhosis.[114]

Circulating immune complexes are major features of various stages of schistosomiasis.[114] Madwar et al.[115] studied adult men with *S. haematobium* or *S. mansoni* infection. Those patients with demonstrable circulating antigen or antigen-antibody complexes had statistically significant decreases in serum C3 levels (about 60% of normal was the lowest) but no changes in C4 or factor B. None of these patients had proteinuria. The chronic hepatosplenic form of schistosomiasis has long been known to be complicated in some patients by proteinuria and other evidence of nephritis. Da Silva et al.[116] studied eight patients with hepatosplenic schistosomiasis (*S. mansoni*) who had no evidence of kidney disease. By light microscopy their glomeruli were normal, but electron microscopy demonstrated mesangial hypertrophy and hyperplasia and electron dense granular deposits on the endothelial basement membrane. Immunofluorescence demonstrated IgG and C3. Retrospective light microscopic studies of patients with hepatosplenic schistosomiasis indicated that approximately 10% had overt glomerulonephritis, incidence being at least twice the incidence in the general autopsy population.[117] Studies by Santoro et al.[118] indicated that immune complexes contained complement components C3, C4, and C1q; levels of circulating immune complexes were inversely correlated with serum C3 levels. In at least one patient, the kidney lesions have been associated with glomerular deposits of immunoglobulins, complement, and adult worm antigens.[119] Patients with *S. haematobium* infection may also have glomerulonephritis, though the role of complement has not been evaluated.[120]

Nephritis has been produced in experimental *S. mansoni* infections in monkeys,[121] mice,[122] and hamsters.[123] In mice,[122] immunoglobulins, complement, and parasitic antigens were demonstrable in the glomeruli. Nephritis has also been produced in experimental infections with *S. japonicum* in hamsters,[123] chimpanzees,[124] and rabbits.[125] Cavallo et al.[124] demonstrated mesangial expansion and subendothelial deposits by electron microscopy but only trace amounts of complement by immunofluorescence. No immunoglobulins or properdin were demonstrable. On the other hand, Jones et al.[125] demonstrated glomerular C3 and immunoglobulin in the 20% of rabbits in their study that developed nephropathy. These rabbits were the ones with persistently high levels of circulating immune complexes and persistently low C3.

One of the most interesting phenomena in the immunology of schistosomiasis is referred to as "concomitant immunity". Infection elicits an immune response that blocks subsequent reinfection but is unable to affect organisms resident in the veins from the original infection. Thus, though no reinfection can occur, adult worms continue to live and produce eggs for many years. No explanation of this phenomenon has been completely satisfactory.[114] It has been suggested that adsorption of host antigens might mask worm surface antigens or worms may express antigenic

Table 3
PUTATIVE COMPLEMENT-DEPENDENT OR COMPLEMENT-ENHANCED HOST DEFENSES AGAINST SCHISTOSOMES

Schistosomal stage	Defense mechanism	Ref.
Cercariae	Lysis	126—128
Schistosomula	Lysis	131—137
	Polymorphonuclear leukocyte phagocytosis	139—141
	Basophil adherence	142
	Eosinophil killing	4,141,143—146

determinants mimicking host antigens. Alternatively, the host immune response may be suppressed by antibodies or by products of the worms.

Upon penetration of skin, cercariae, which previously have had a surface coat of neutral polysaccharides, lose this neutral coat and expose an acidic glycan layer. The epithelium of these schistosomula then proliferates. Older schistosomula and adult worms have a very active surface epithelium with continual shedding and replacement of surface components. One of the surface components may be sialic acid.[96]

The primary interaction between the complement system and schistosomes is during the first hours and days following skin penetration by the cercariae and their conversion and migration as schistosomula (Tables 3 and 4). Culbertson[126] demonstrated in 1936 an activity in vertebrate serum that was lethal for cercariae of various species in vitro. He furthermore demonstrated that complement was consumed in the process and the cercaricidal activity of serum was heat labile, inactivated by dried cercariae or bacteria at room temperature for 3 hr, and destroyed by known complement inactivating agents including yeast cells and ammonium hydroxide. Standen[127] studied specifically the cercariae of *S. mansoni* and their reactions in normal sera from various species. Effects of serum ranged from killing of all cercariae by 3 hr to a reaction characterized by rapid onset of immotility followed by tail loss. (One of the characteristics of cercarial penetration and conversion to schistosomula is the loss of the cercarial tail.) In this last reaction, though bodies of cercariae began to die as early as 3 hr, many cercariae survived up to 24 hr. This activity of normal serum was heat labile. A different reaction was observed with immune sera. Precipitates, visible by light microscopy, developed around cercariae, which became immotile within minutes and adhered to glass. This activity was not strictly cercaricidal and the tails of the cercariae were not lost. The activity of immune serum was not heat labile to any great extent.

Machado et al.[128] further defined the importance of complement in the cercaricidal reaction. Incubation of cercariae in normal sera for 60 min at 37°C consumed hemolytic complement activity; C3 was immunoelectrophoretically converted as well. In human serum, but not in guinea pig serum, C4 was also consumed by this reaction, whereas C1 and C2 were not affected in human or guinea pig serum. Furthermore, the cercariae developed surface C3 during this reaction as demonstrated by immunofluorescence. Infectivity of cercariae was markedly decreased and cercaricidal ability of serum parallelled hemolytic complement levels. The cercaricidal activity of serum was not affected by C4 deficiency or EGTA chelation but was markedly diminished by heating at 50°C for 30 min. Cercaricidal activity of serum was completely abolished by heating at 56°C for 30 min; prior incubation with zymosan,

Table 4
CERCARIACIDAL AND SCHISTOSOMULACIDAL ACTIVITY OF NORMAL AND IMMUNE SERA FROM VARIOUS SPECIES

Schistosomal stage	Serum source	Normal serum			Immune serum		Ref.
		Killing	C^a	ACP^b	Killing	C^a	
Cercariae	Human	+	+	+	ND^c	ND	128
		−	ND	ND	ND	ND	129
	Guinea pig	+	+	+	ND	ND	128
		+	+	ND	+	−	127
	Other vertebrates^d	+	+	ND	ND	ND	126^e
		+	+	ND	+	−	127
Schistosomula	Human	−	ND	ND	ND	ND	130
		+	+	+	ND	ND	133
	Monkey	−	ND	ND	ND	ND	131
		+	+	+	ND	ND	133
	Rat	−	ND	ND	ND	ND	131
	Guinea pig	+	ND	ND	+	ND	132
		+	+	+	ND	ND	133
	Chicken	+	+	+	ND	ND	133

^{a} Killing required complement.
^{b} Serum killing could be demonstrated when only alternative pathway functional.
^{c} Not done.
^{d} Snake, frog, rabbit, rat, mouse, cat, dog, horse, sheep, pig.
^{e} Used *Schistosoma douthitti* rather than *S. mansoni*.

inulin, cobra venom factor, or hydrazine; chelation with EDTA; or depletion of properdin. These results suggested that the alternative complement pathway played a role in host defenses against the invading cercariae. On the other hand, Greenblatt et al.[129] suggested a somewhat different interpretation of these events. These authors were not able to demonstrate cercaricidal activity of human serum even in concentrations of 50%. They pointed out that the evolution of cercariae to schistosomula involved complex events that included tail loss, loss of an ability to survive in water, various changes in glands of the larvae, ability to grow in tissue culture, and changes in the outer membrane. They suggested that serum factors were important in this developmental transformation of the cercariae rather than in any host defenses. They corroborated the observation that cercariae incubated in normal serum acquired surface-bound C3. This C3 binding was inhibited by heat or EDTA chelation but not by EGTA chelation. Complement deposition on the cercariae was accompanied by tail loss. The tail loss effect of serum could be inhibited by cobra venom factor; preincubation of the serum with sepharose, zymosan, or antiserum to C3; or heat inactivation at 50°C for 30 min. The effect was normal in C2-, C5-, C7-, and C8-deficient human sera; C4-deficient guinea pig serum; C5-deficient mouse serum; C6-deficient rabbit serum; hypogammaglobulinemic serum; and in serum absorbed with cercariae. An explanation for the discrepancy between the observation of Machado et al.[128] and Greenblatt et al.[129] is not apparent.

The ability of normal serum to kill schistosomula has been studied with varying results. Kusel[130] demonstrated agglutination of schistosomula but not killing over several hours incubation with normal human serum. Likewise Sher et al.[131] could not demonstrate lethal activity of rat or rhesus monkey serum. On the other hand, Tavares et al.[132] demonstrated killing by normal guinea pig serum, though that killing was not so great as by immune serum. And Santoro et al.[133] demonstrated schistosomulacidal activity over a 4-hr period in vitro with normal chicken serum, guinea pig serum, human serum, and monkey serum. Lethal activity in these normal sera was apparently due to alternative pathway activation and could be blocked by depletion of factor B or factor D, by zymosan treatment of serum, or by dilution of serum. On the other hand, killing was normal in C4-deficient guinea pig serum, C2-deficient human serum, and agammaglobulinemic human serum. They furthermore showed that factor B and functional alternative pathway activity were both consumed in the process.

Immune sera from various species including man have uniformly been shown to kill schistosomula in a complement-dependent fashion.[131-137] Though evaluated in only one study, the complement activation by immune serum proceeded primarily through the classical pathway.[132] Kassis et al.[137] also demonstrated that this activity occurred in vivo as well as in vitro. These studies generally used schistosomula cultured in vitro. Clegg and Smithers[134] demonstrated that, if the in vitro cultures included monkey serum or if the schistosomula were grown in vivo in mice, schistosomula were no longer susceptible to the lethal effect of antibody and complement. Furthermore, schistosomula obtained from lungs early in infection were sensitive to the activity of normal serum, whereas schistosomula obtained from the lung 4 days after infection were resistant.[133] Schistosomula acquire some form of protection after growth in the host over several days; if antibody and complement are to have any effect they must do so early in infection. Two studies[131,135] could not demonstrate protection to be induced by immunization, even though immune serum had in vitro schistosomulacidal activity. Tavares et al.,[132] however, found that immune mice had about half the worm burden (i.e., the number of living adult worms in venous plexuses) of nonimmune mice; these data indicated some decrease in the number of schistosomula able to develop into adult worms. Cobra venom factor partially or completely inhibited that protection.

Other investigations have suggested that, in addition to antibody and complement, leukocytes are also important in killing of schistosomula. Gazinelli et al.[138] demonstrated that cercarial extracts incubated with plasma or serum generated an anaphylatoxin through the complement system. Thus, it is possible that leukocytes accumulate through a complement-mediated mechanism. Studies with rat,[139] guinea pig,[140] and human[141] systems indicated that immunoglobulin and complement were required for granulocyte killing of schistosomula of *S. mansoni*. Even when antiserum alone was schistosomulacidal, addition of polymorphonuclear leukocytes enhanced killing.[140] Anwar et al.[141] suggested also that complement alone, activated via the alternative pathway, could mediate killing by granulocytes and mononuclear leukocytes.

Sher[142] noted that mast cell (basophil) infiltration was often seen at sites of Helminthic infections and that schistosomula incubated with immune or normal rat serum and mast cells became coated with adherent mast cells. The activity of immune serum was dependent on complement. The activity of nonimmune serum was also dependent on complement, primarily through the alternative pathway as indicated by failure to inhibit with EGTA. He further substantiated the last point by incubating schistosomula with normal, nonimmune rat serum in the absence of mast cells and demonstrating subsequently surface C3 but not immunoglobulin by immunofluorescence.

As noted earlier, eosinophilia is a characteristic feature of Helminthic infections. The first identification of a "beneficial" role for this eosinophilia came from observations that eosinophils killed schistosomula in an antibody-dependent fashion.[4] Subsequent studies have also shown an important role for the complement system.

Ottesen et al.[143] found that immune human serum mediated adherence of eosinophils as well as polymorphonuclear leukocytes to schistosomula. This serum activity was heat labile and abrogated by zymosan treatment. Nonimmune and C2-deficient sera mediated adherence almost as well as immune serum. They concluded that the alternative complement pathway, with or without antibody, mediated binding of eosinophils and polymorphonuclear leukocytes to schistosomula. Ramalho-Pinto et al.[144] subsequently showed that normal and immune rat sera also mediated killing of schistosomula by eosinophils. Heat inactivation completely destroyed the effect of nonimmune serum but had no demonstrable effect on adherence or killing by eosinophils in immune rat serum. The effect of nonimmune serum appeared to be mediated through the alternative pathway. However, they looked further at the quantitative requirements for eosinophils to kill schistosomula in diluted serum. If heat-inactivated immune serum were used, 3500 eosinophils were required to kill each schistosomulum. If diluted nonimmune rat serum were used, only 900 eosinophils were required for killing. These data are reminiscent of the studies indicating the marked effect of complement in enhancing the efficiency of phagocyte interactions with bacteria. McLaren and colleagues[145,146] have further evaluated the roles of eosinophils and various serum factors. In kinetic analyses, immune rat serum was more efficient; essentially 100% of schistosomula were dead by 10 to 20 hr, whereas 30 to 40 hr were required for the same effect in normal serum. Activity of nonimmune serum was totally dependent on complement; the effect of immune serum was reduced by about half by heat inactivation. Anwar et al.[141] also examined the participation of complement in killing of schistosomula by eosinophils and suggested that complement was an absolute requirement.

Adult schistosomes in the venous system exist in symbiosis with the host. One of the suspected mechanisms is incorporation of host proteins. Kabil[147] examined worms of various ages with immunofluorescence and demonstrated progressive accumulation of C3 antigens on the surface of the schistosomes. By 9 to 13 weeks of age female schistosomes were strongly positive by immunofluorescence with antiserum to C3. Interestingly, male schistosomes never developed surface C3.

Butterworth[148] has reviewed the various effector mechanisms against schistosomula in vitro and has questioned the ability to generalize these observations to the in vivo situation. Conditions in vitro are suboptimum for the parasite and such mechanisms may be blocked in vivo. Furthermore, in view of the considerable time required in vitro for some activities to be demonstrable (20 to 40 hr), it is questionable how relevant these mechanisms are in vivo where the host defense systems may have only limited access to the parasite. Nevertheless, it seems likely that one of the more important host defense mechanisms in limiting the number of schistosomula that are able to mature and develop into adult schistosomes is the interaction of antibody, complement, and granulocytes.

C. Nematodes (Round Worms)

The nematodes are relatively simple compared to other Helminths. They have a smooth external cuticle formed by secretions of the underlying cells.[149] Three patterns of parasitism occur: nematodes may live their entire lives in the intestines of the host; others have a tissue phase associated with larval migration, but the major site of infection so far as the worm is concerned is the intestinal tract; and others, such as *Trichinella spiralis*, have primarily a tissue phase. Those nematodes that exist almost exclusively in the intestinal tract have limited chance for interaction with the complement system. In studies of *Nippostrongylus brasiliensis* infection in rats, Jones and Ogilvie[150] could demonstrate no participation by complement in expulsion of worms from the guts of immune rats (though their evidence suggested some role for the immune system in this event).

Ascaris infections are characterized by both tissue and intestinal phases. Larvae penetrate the intestinal mucosa and migrate via the bloodstream to the lungs, then through the respiratory tract to the intestinal tract where they develop into mature worms. The primary interactions of the host defense systems therefore are with the tissue phases of the infection. Ziprin and Jeska[151] studied interactions of larvae of *Ascaris suum* with the host immune system. Peritoneal exudate cells, in the presence of peritoneal fluid, adhered to larvae. Heat inactivation of the exudate fluid largely abrogated this effect; activity was restored by addition of guinea pig serum. Participation of complement was further suggested by blockage of this reaction with EDTA chelation. Leventhal and Soulsby[152] demonstrated binding of C3 by infective-stage larvae. Animals depleted of complement with cobra venom factor or congenitally deficient in C4 developed prominent pneumonitis and unusually large numbers of organisms in the lung after experimental infection with *A. suum*.[153] Thus, as is the case with other parasitic diseases characterized by larval penetration, complement seems to play some role in early defenses against dissemination and progressive infection.

Larvae of *T. spiralis* disseminate widely and encyst in various tissues, particularly muscle. Stankiewicz and Jeska[154] demonstrated that normal peritoneal exudate fluid mediated adherence of leukocytes to these larvae. The factor in the exudate fluid was heat labile, though a heat stable factor could be induced by immunity. Stankiewicz[155] suggested that the alternative complement pathway was the primary mediating factor in normal rat or guinea pig serum, which could be substituted for the peritoneal exudate fluid.

Tropical pulmonary eosinophilia is a syndrome of recurrent pneumonitis and peripheral eosinophilia. The etiology is thought to be passage of parasites through the lung with a local immunological reaction. It is likely that various organisms are involved, but some of the microfilaria are especially important. One study has been published of the serum complement levels in 19 patients during such disease.[156] In general, C3 levels were elevated. C4 levels were occasionally abnormally low. All low C4 values occurred among patients with primary disease.

REFERENCES

1. **Santoro, F., Bernal, J., and Capron, A.,** Complement activation by parasites. A review, *Acta Trop.*, 36, 5, 1979.
2. **Mahmoud, A. A. F.,** Protozoa, in *Immunological Diseases*, Samter, M., Ed., Little, Brown, Boston, 1978, 738.
3. **Ogilvie, B. M. and Wilson, R. J. M.,** Evasion of the immune response by parasites, *Br. Med. Bull.*, 32, 177, 1976.
4. **Warren, K. S.,** Worms, in *Immunological Diseases*, Samter, M., Ed., Little, Brown, Boston, 1978, 718.
5. **Anwar, A. R. E. and Kay, A. B.,** Membrane receptors for IgG and complement (C4, C3b and C3d) on human eosinophils and neutrophils and their relation to eosinophilia, *J. Immunol.*, 119, 976, 1977.

6. **Brown, I. N.,** Immunological aspects of malaria infection, *Adv. Immunol.,* 11, 267, 1969.
7. **Neva, F. A., Sheagren, J. N., Shulman, N. R., and Canfield, C. J.,** Malaria: host defense mechanisms and complications, *Ann. Intern. Med.,* 73, 295, 1970.
8. **Cohen, S. and Butcher, G. A.,** Properties of protective malarial antibody, *Immunology,* 19, 369, 1970.
9. **Cohen, S. and Butcher, G. A.,** Serum antibody in acquired malarial immunity, *Trans. R. Soc. Trop. Med. Hyg.,* 65, 125, 1971.
10. **Diggs, C. L., Shin, H., Briggs, N. T., Laudenslayer, K., and Weber, R.,** Antibody mediated immunity to *Plasmodium berghei* independent of the third component of complement, in *Proc. Helm. Soc. Wash.,* 39, 456, 1972.
11. **Williams, A. I. O., Rosen, F. S., and Hoff, R.,** Role of complement components in the susceptibility to *Plasmodium berghei* infection among inbred strains of mice, *Ann. Trop. Med. Parasitol.,* 69, 179, 1975.
12. **Roy, A. N. and Mukerjee, S.,** Some observations on the complement in the serum of monkeys during infection with *Plasmodium knowlesi, Ann. Biochem. Exp. Med.,* 2, 245, 1942.
13. **Dulaney, A. D.,** The complement content of human sera with especial reference to malaria, *J. Clin. Invest.,* 27, 320, 1948.
14. **Fogel, B. J., von Doenhoff, A. E., Jr., Cooper, N. R., and Fife, E. H., Jr.,** Complement in acute experimental malaria. I. Total hemolytic activity, *Milit. Med.,* 131 (Suppl.), 1173, 1966.
15. **Cooper, N. R. and Fogel, B. J.,** Complement in acute experimental malaria. II. Alterations in the components of complement, *Milit. Med.,* 131 (Suppl.), 1180, 1966.
16. **Topley, E., Knight, R., and Woodruff, A. W.,** The direct antiglobulin test and immunoconglutinin titres in patients with malaria, *Trans. R. Soc. Trop. Med. Hyg.,* 67, 51, 1973.
17. **Greenwood, B. M. and Brueton, M. J.,** Complement activation in children with acute malaria, *Clin. Exp. Immunol.,* 18, 267, 1974.
18. **Srichaikul, T., Puwasatien, P., Puwasatien, P., Karnjanajetanee, J., and Bokisch, V. A.,** Complement changes and disseminated intravascular coagulation in *Plasmodium falciparum* malaria, *Lancet,* 1, 770, 1975.
19. **Ree, G. H.,** Complement and malaria, *Ann. Trop. Med. Parasitol.,* 70, 247, 1976.
20. **Petchclai, B., Chutanondh, R., Hiranras, S., and Benjapongs, W.,** Activation of classical and alternate complement pathways in acute falciparum malaria, *J. Med. Assoc. Thailand,* 60, 174, 1977.
21. **Rose, J. E., Amerault, T. E., Roby, T. O., and Martin, W. H.,** Serum levels of conglutinin, complement, and immunoconglutinin in cattle infected with *Anaplasma marginale, Am. J. Vet. Res.,* 39, 791, 1978.
22. **Neva, F. A., Howard, W. A., Glew, R. H., Krotoski, W. A., Gam, A. A., Collins, W. E., Atkinson, J. P., and Frank, M. M.,** Relationship of serum complement levels to events of the malarial paroxysm, *J. Clin. Invest.,* 54, 451, 1974.
23. **Glew, R. H., Atkinson, J. P., Frank, M. M., Collins, W. E., and Neva, F. A.,** Serum complement and immunity in experimental simian malaria. I. Cyclical alterations in C4 related to schizont rupture, *J. Infect. Dis.,* 131, 17, 1975.
24. **Atkinson, J. P., Glew, R. H., Neva, F. A., and Frank, M. M.,** Serum complement and immunity in experimental simian malaria. II. Preferential activation of early components and failure of depletion of late components to inhibit protective immunity, *J. Infect. Dis.,* 131, 26, 1975.
25. **Connal, A.,** Auto-erythrophagocytosis in protozoal diseases, *J. Pathol. Bacteriol.,* 16, 502, 1912.
26. **Todorovic, R., Ferris, D. H., and Ristic, M.,** Antigens of *Plasmodium gallinaceum.* II. Immunoserologic characterization of explasmodial antigens and their antibodies, *Am. J. Trop. Med. Hyg.,* 17, 695, 1968.
27. **Ward, P. A. and Conran, P. B.,** Immunopathologic studies of simian malaria, *Milit. Med.,* 131 (Suppl.), 1225, 1966.
28. **Soothill, J. F. and Hendrickse, R. G.,** Some immunological studies of the nephrotic syndrome of Nigerian children, *Lancet,* 2, 629, 1967.
29. **Allison, A. C., Houba, V., Hendrickse, R. G., de Petris, S., Edington, G. M., and Adeniyi, A.,** Immune complexes in the nephrotic syndrome of African children, *Lancet,* L, 1232, 1969.
30. **Ward, P. A. and Kibukamusoke, J. W.,** Evidence for soluble immune complexes in the pathogenesis of the glomerulonephritis of quartan malaria, *Lancet,* 1, 283, 1969.
31. **Ward, P. A. and Conran, P. B.,** Immunopathology of renal complications in simian malaria and human quartan malaria, *Milit. Med.,* 134, 1228, 1969.
32. **June, C. H., Contreras, C. E., Perrin, L. H., Lambert, P. H., and Miescher, P. A.,** Circulating and tissue-bound immune complex formation in murine malaria, *J. Immunol.,* 122, 2154, 1979.
33. **Bhamarapravati, N., Boonpucknavig, S., Boonpucknavig, V., and Yaemboonruang, C.,** Glomerular changes in acute *Plasmodium falciparum* infection, *Arch. Pathol.,* 96, 289, 1973.

34. **Chapman, W. E. and Ward, P. A.**, *Babesia rodhaini:* requirement for complement for penetration of human erythrocytes, *Science*, 196, 67, 1977.
35. **Chapman, W. E. and Ward, P. A.**, Changes in C3 metabolism during protozoan infection (*Babesia rodhaini*) in rats, *J. Immunol.*, 116, 1284, 1976.
36. **Chapman, W. E. and Ward, P. A.**, The complement profile in babesiosis, *J. Immunol.*, 117, 935, 1976.
37. **Annable, C. R. and Ward, P. A.**, Immunopathology of the renal complications of babesiosis, *J. Immunol.*, 112, 1, 1974.
38. **Hawking, F.**, The *Trypanosomidae*, in *Tropical Medicine*, Hunter, G. W., III, Swartzwelder, J. C., and Clyde, D. F., Eds., W. B. Saunders, Philadelphia, 1976, 408.
39. **Hawking, F.**, African *trypanosomidae*, in *Tropical Medicine*, Hunter, G. W., III, Swartzwelder, J. C., and Clyde, D. F., Eds., W. B. Saunders, Philadelphia, 1976, 430.
40. **Eyckmans, L.**, Trypanosoma species (sleeping sickness and Chagas' disease), in *Principles and Practice of Infectious Diseases*, Mandell, G. L., Douglas, R. G., Jr., and Bennett, J. E., Eds., John Wiley & Sons, New York, 1979, 2118.
41. **Ingram, D. G. and Soltys, M. A.**, Immunity in trypanosomiasis. IV. Immuno-conglutinin in animals infected with *Trypanosoma brucei*, *Parasitology*, 50, 231, 1960.
42. **Barrett-Connor, E., Ugoretz, R. J., and Braude, A. I.**, Disseminated intravascular coagulation in trypanosomiasis, *Arch. Intern. Med.*, 131, 574, 1973.
43. **Greenwood, B. M. and Whittle, H. C.**, Complement activation in patients with Gambian sleeping sickness, *Clin. Exp. Immunol.*, 24, 133, 1976.
44. **Assoku, R. K. G., Tizard, I. R., and Nielsen, K. H.**, Free fatty acids, complement activation, and polyclonal B-cell stimulation as factors in the immunopathogenesis of African trypanosomiasis, *Lancet*, 2, 956, 1977.
45. **Musoke, A. J. and Barbet, A. F.**, Activation of complement by variant-specific surface antigen of *Trypanosoma brucei*, *Nature (London)*, 270, 438, 1977.
46. **Tizard, I., Nielsen, K. H., Seed, J. R., and Hall, J. E.**, Biologically active products from African trypanosomes, *Microbiol. Rev.*, 42, 661, 1978.
47. **Murray, M.**, The pathology of African trypanosomiasis, in *Progress in Immunology II. Vol. 4: Clinical Aspects I.*, Brent, L. and Holborow, J., Eds., North-Holland, Amsterdam, 1974, 181.
48. **Woodruff, A. W., Ziegler, J. L., Hathaway, A., and Gwata, T.**, Anemia in African trypanosomiasis and "big spleen disease" in Uganda, *Trans. R. Soc. Trop. Med. Hyg.*, 67, 329, 1973.
49. **Nagle, R. B., Ward, P. A., Lindsley, H. B., Sadun, E. H., Johnson, A. J., Berkaw, R. E., and Hildebrandt, P. K.**, Experimental infections with African trypanosomes. VI. Glomerulonephritis involving the alternate pathway of complement activation, *Am. J. Trop. Med. Hyg.*, 23, 15, 1974.
50. **Yorke, W., Adams, A. R. D., and Murgatroyd, F.**, Studies in chemotherapy. II. The action *in vitro* of normal human serum on the pathogenic trypanosomes, and its significance, *Ann. Trop. Med. Parasitol.*, 24, 115, 1930.
51. **Corson, J. F.**, Experiments on the transmission of *Trypanosoma brucei* and *Trypanosoma rhodesiense* to man, *Ann. Trop. Med. Parasitol.*, 26, 109, 1932.
52. **Adams, A. R. D.**, A record of an investigation into the action of sera on the trypanosomes pathogenic to man, *Ann. Trop. Med. Parasitol.*, 27, 309, 1933.
53. **Aaronovitch, S. and Terry, R. J.**, The trypanolytic factor in normal human serum, *Trans. Roy. Soc. Trop. Med. Hyg.*, 66, 344, 1972.
54. **Diggs, C., Flemmings, B., Dillon, J., Snodgrass, R., Campbell, G., and Esser, K.**, Immune serum-mediated cytotoxicity against *Trypanosoma rhodesiense*, *J. Immunol.*, 116, 1005, 1976.
55. **Balber, A. E., Bangs, J. D., Jones, S. M., and Proia, R. L.**, Inactivation or elimination of potentially trypanolytic, complement-activating immune complexes by pathogenic trypanosomes, *Infect. Immun.*, 24, 617, 1979.
56. **Flemmings, B. and Diggs, C.**, Antibody-dependent cytotoxicity against *Trypanosoma rhodesiense* mediated through an alternative complement pathway, *Infect. Immun.*, 19, 928, 1978.
57. **Kobayashi, A. and Tizard, I. R.**, The response to *Trypanosoma congolense* infection in calves. Determination of immunoglobulins IgG1, IgG2, IgM and C3 levels and the complement fixing antibody titres during the course of infection, *Tropenmed. Parasitol.*, 27, 411, 1976.
58. **Nielsen, K., Sheppard, J., Holmes, W., and Tizard, I.**, Experimental bovine trypanosomiasis. Changes in the catabolism of serum immunoglobulins and complement components in infected cattle, *Immunology*, 35, 811, 1978.
59. **Nielsen, K., Sheppard, J., Holmes, W., and Tizard, I.**, Experimental bovine trypanosomiasis. Changes in serum immunoglobulins, complement and complement components in infected animals, *Immunology*, 35, 817, 1978.

60. **Kobayashi, A., Tizard, I. R., and Woo, P. T. K.,** Studies on the anemia in experimental African trypanosomiasis. II. The pathogenesis of the anemia in calves infected with *Trypanosoma congolense, Am. J. Trop. Med. Hyg.,* 25, 401, 1976.
61. **Nielsen, K. and Sheppard, J.,** Activation of complement by trypanosomes, *Experientia,* 33, 769, 1977.
62. **Nielsen, K., Sheppard, J., Tizard, I., and Holmes, W.,** Direct activation of complement by trypanosomes, *J. Parasitol.,* 64, 544, 1978.
63. **Nielsen, K., Sheppard, J., Tizard, I., and Holmes, W.,** *Trypanosoma lewisi:* characterization of complement-activating components, *Exp. Parasitol.,* 43, 153, 1977.
64. **Nielsen, K., Sheppard, J., Holmes, W., and Tizard, I.,** Increased susceptibility of *Trypanosoma lewisi* infected, or decomplemented rats to *Salmonella typhimurium, Experientia,* 34, 118, 1978.
65. **Lange, D. E. and Lysenko, M. G.** *In vitro* phagocytosis of *Trypanosoma lewisi* by rat exudative cells, *Exp. Parasitol.,* 10, 39, 1960.
66. **Jarvinen, J. A. and Dalmasso, A. P.,** Complement in experimental *Trypanosoma lewisi* infections of rats, *Infect. Immun.,* 14, 894, 1976.
67. **Dusanic, D. G.,** *Trypanosoma musculi* infections in complement-deficient mice, *Exp. Parasitol.,* 37, 205, 1975.
68. **Jarvinen, J. A. and Dalmasso, A. P.,** *Trypanosoma musculi* infections in normocomplementemic, C5-deficient, and C3-depleted mice, *Infect. Immun.,* 16, 557, 1977.
69. **Jarvinen, J. A. and Dalmasso, A. P.,** *Trypanosoma musculi:* immunologic features of the anemia in infected mice, *Exp. Parasitol.,* 43, 203, 1977.
70. **Taliaferro, W. H.,** Ablastic and trypanocidal antibodies against *Trypanosoma duttoni, J. Immunol.,* 35, 303, 1938.
71. **Kierszenbaum, F. and Weinman, D.,** Antibody-independent activation of the alternative complement pathway in human serum by parasitic cells, *Immunology,* 32, 245, 1977.
72. **Arean, V. M.,** American trypanosomiasis, in *Tropical Medicine,* Hunter, G. W., III, Swartzwelder, J. C., and Clyde, D. F., Eds., W. B. Saunders, Philadelphia, 1976, 440.
73. **Cunningham, D. S., Craig, W. H., and Kuhn, R. E.,** Reduction of complement levels in mice infected with *Trypanosoma cruzi, J. Parasitol.,* 64, 1044, 1978.
74. **Budzko, D. B., Pizzimenti, M. C., and Kierszenbaum, F.,** Effects of complement depletion in experimental Chagas' disease: immune lysis of virulent blood forms of *Trypanosoma cruzi, Infect. Immun.,* 11, 86, 1975.
75. **Denison, N.,** Immunologic studies on experimental *Trypanosoma cruzi* infections: lysins in blood of infected rats, in *Proc. Soc. Exp. Biol. Med.,* 52, 26, 1943.
76. **Anziano, D. F., Dalmasso, A. P., Lelchuk, R., and Vasquez, C.,** Role of complement in immune lysis of *Trypanosoma cruzi, Infect. Immun.,* 6, 860, 1972.
77. **Nogueira, N., Bianco, C., and Cohn, Z.,** Studies on the selective lysis and purification of *Trypanosoma cruzi, J. Exp. Med.,* 142, 224, 1975.
78. **Kierszenbaum, F., Ivanyi, J., and Budzko, D. B.,** Mechanisms of natural resistance to trypanosomal infection. Role of complement in avian resistance to *Trypanosoma cruzi* infection, *Immunology,* 30, 1, 1976.
79. **Kierszenbaum, F.,** Cross-reactivity of lytic antibodies against blood forms of *Trypanosoma cruzi, J. Parasitol.,* 62, 134, 1976.
80. **Biagi, F.,** Leishmaniasis-Introduction, in *Tropical Medicine,* Hunter, G. W., III, Swartzwelder, J. C., and Clyde, D. F., Eds., W. B. Saunders, Philadelphia, 1976, 411.
81. **Rocha, H.,** Leishmania species (kala-azar), in *Principles and Practice of Infectious Diseases,* Mandell, G. L., Douglas, R. G., Jr., and Bennett, J. E., Eds., John Wiley & Sons, New York, 1979, 2110.
82. **Mauel, J. and Behin, R.,** Cell-mediated and humoral immunity to protozoan infections (with special reference to leishmaniasis), *Transplant. Rev.,* 19, 121, 1974.
83. **Rezai, H. R., Ardehali, S. M., Amirhakimi, G., and Kharazmi, A.,** Immunological features of kala-azar, *Am. J. Trop. Med. Hyg.,* 27, 1079, 1978.
84. **Woodruff, A. W., Topley, E., Knight, R., and Downie, C. G. B.,** The anaemia of kala azar, *Br. J. Haematol.,* 22, 319, 1972.
85. **Adler, S.,** Attempts to transmit visceral leishmaniasis to man. Remarks on the histopathology of leishmaniasis, *Trans. R. Soc. Trop. Med. Hyg.,* 33, 419, 1940.
86. **Ulrich, M., Trujillo Ortiz, D., and Convit, J.,** The effect of fresh serum on the leptomonads of *Leishmania.* I. Preliminary report, *Trans. R. Soc. Trop. Med. Hyg.,* 62, 825, 1968.
87. **Schmunis, G. A. and Herman, R.,** Characteristics of so-called natural antibodies in various normal sera against culture forms of *Leishmania, J. Parasitol.,* 56, 889, 1970.

88. **Lainson, R., and Strangways-Dixon, J.,** *Leishmania mexicana:* the epidemiology of dermal leishmaniasis in British Honduras, *Trans. R. Soc. Trop. Med. Hyg.*, 57, 242, 1963.
89. **Ganguly, N. K., Mahajan, R. C., Datta, D. V., Sharma, S., Chhuttani, P. N., and Gupta, A. K.,** Immunoglobulin and complement levels in cases of invasive amoebiasis, *Indian J. Med. Res.*, 67, 221, 1978.
90. **Ortiz-Ortiz, L., Capin, R., Capin, N. R., Sepulveda, B., and Zamacona, G.,** Activation of the alternative pathway of complement by *Entamoeba histolytica*, *Clin. Exp. Immunol.*, 34, 10, 1978.
91. **Anderson, S.,** *Toxoplasma gondii*, in *Principles and Practice of Infectious Diseases*, Mandell, G. L., Douglas, R. G., Jr., and Bennett, J. E., Eds., John Wiley & Sons, New York, 1979, 2127.
92. **Araujo, F. G., Rosenberg, L. T., and Remington, J. S.,** Experimental *Toxoplasma gondii* infection in mice: the role of the fifth component of complement, *Proc. Soc. Exp. Biol. Med.*, 149, 800, 1975.
93. **Sabin, A. B. and Feldman, H. A.,** Dyes as microchemical indicators of a new immunity phenomenon affecting a protozoan parasite (toxoplasma), *Science*, 108, 660, 1948.
94. **Gronroos, P.,** The action of properdin on *Toxoplasma gondii*. A preliminary report, *Ann. Med. Exp. Biol. Fenn.*, 33, 310, 1955.
95. **Feldman, H. A.,** The relationship of *Toxoplasma* antibody activator to the serum-properdin system, *Ann. N.Y. Acad. Sci.*, 66, 263, 1956.
96. **Lumsden, R. D.,** Parasitological review: surface ultrastructure and cytochemistry of parasitic helminths, *Exp. Parasitol.*, 37, 267, 1975.
97. **Jones, T. C.,** Cestodes (Tapeworms), in *Principles and Practice of Infectious Diseases*, Mandell, G. L., Douglas, R. G., Jr., and Bennett, J. E., Eds., John Wiley & Sons, New York, 1979, 2183.
98. **Murrell, K. D.,** The effect of antibody on the permeability control of larval *Taenia taeniaeformis*, *J. Parasitol.*, 57, 875, 1971.
99. **Heath, D. D.,** Resistance to *Taenia pisiformis* larvae in rabbits. II. Temporal relationships and the development phase affected, *Int. J. Parasitol.*, 3, 491, 1973.
100. **Musoke, A. J. and Williams, J. F.,** The immunological response of the rat to infection with *Taenia taeniaeformis*. V. Sequence of appearance of protective immunoglobulins and the mechanism of action of 7Sγ2a antibodies, *Immunology*, 29, 855, 1975.
101. **Mitchell, G. F., Goding, J. W., and Rickard, M. D.,** Studies on immune responses to larval cestodes in mice. Increased susceptibility of certain mouse strains and hypothymic mice to *Taenia taeniaeformis* and analysis of passive transfer of resistance with serum, *Aus. J. Exp. Biol. Med. Sci.*, 55, 165, 1977.
102. **Hustead, S. T. and Williams, J. F.,** Permeability studies on taeniid metacestodes. II. Antibody-mediated effects on membrane permeability in larvae of *Taenia taeniaeformis* and *Taenia crassiceps*, *J. Parasitol.*, 63, 322, 1977.
103. **Hammerberg, B. and Williams, J. F.,** Interaction between *Taenia taeniaeformis* and the complement system, *J. Immunol.*, 120, 1033, 1978.
104. **Hammerberg, B. and Williams, J. F.,** Physicochemical characterization of complement-interacting factors from *Taenia taeniaeformis*, *J. Immunol.*, 120, 1039, 1978.
105. **Heyneman, D. and Welsh, J. F.,** Action of homologous antiserum *in vitro* against life cycle stages of *Hymenolepis nana*, the dwarf mouse tapeworm, *Exp. Parasitol.*, 8, 119, 1959.
106. **Befus, A. D.,** *Hymenolepis diminuta* and *H. microstoma:* mouse immunoglobulins binding to the tegumental surface, *Exp. Parasitol.*, 41, 242, 1977.
107. **Threadgold, L. T. and Befus, A. D.,** *Hymenolepis diminuta:* ultrastructural localization of immunoglobulin-binding sites on the tegument, *Exp. Parasitol.*, 43, 169, 1977.
108. **Herd, R. P.,** The cestocidal effect of complement in normal and immune sera *in vitro*, *Parasitology*, 72, 325, 1976.
109. **Kassis, A. I. and Tanner, C. E.,** The role of complement in hydatid disease: *in vitro* studies, *Int. J. Parasitol.*, 6, 25, 1976.
110. **Rickard, M. D., Mackinlay, L. M., Kane, G. J., Matossian, R. M., and Smyth, J. D.,** Studies on the mechanism of lysis of *Echinococcus granulosus* protoscoleces incubated in normal serum, *J. Helminthol.*, 51, 221, 1977.
111. **Kassis, A. I. and Tanner, C. E.,** Host serum proteins in *Echinococcus multilocularis:* complement activation via the classical pathway, *Immunology*, 33, 1, 1977.
112. **Kassis, A. I., Goh, S. L., and Tanner, C. E.,** Lesions induced by complement *in vitro* on the protoscoleces of *Echinococcus multilocularis:* a study by electron microscopy, *Int. J. Parasitol.*, 6, 199, 1976.
113. **Kassis, A. I. and Tanner, C. E.,** *Echinococcus multilocularis:* complement's role *in vivo* in hydatid disease. *Exp. Parasitol.*, 43, 390, 1977.
114. **Phillips, S. M. and Colley, D. G.,** Immunologic aspects of host responses to schistosomiasis: resistance, immunopathology, and eosinophil involvement, *Progr. Allergy*, 24, 49, 1978.

115. **Madwar, M. A., O'Shea, J. M., Skelton, J. A., and Soothill, J. F.,** Complement components and immunoglobulins in patients with schistosomiasis, *Clin. Exp. Immunol.*, 34, 354, 1978.
116. **da Silva, L. C., de Brito, T., Camargo, M. E., de Boni, D. R., Lopes, J. D., and Gunji, J.,** Kidney biopsy in the hepatosplenic form of infection with *Schistosoma mansoni* in man, *Bull. World Health Organization*, 42, 907, 1970.
117. **Andrade, Z. A., Andrade, S. G., and Sadigursky, M.,** Renal changes in patients with hepatosplenic schistosomiasis, *Am. J. Trop. Med. Hyg.*, 20, 77, 1971.
118. **Santoro, F., Bout, D., Camus, D., and Capron, A.,** Immune complexes in schistosomiasis. IV. C3, C4 and C1q characterization and correlation between C3 in serum and circulating IC levels, *Rev. Inst. Med. Trop. Sao Paulo*, 19, 39, 1977.
119. **Falcao, H. A. and Gould, D. B.,** Immune complex nephropathy in schistosomiasis, *Ann. Intern. Med.*, 83, 148, 1975.
120. **Sabbour, M. S., El-Said, W., and Abou-Gabal, I.,** A clinical and pathological study of schistosomal nephritis, *Bull. W.H.O.*, 47, 549, 1972.
121. **de Brito, T., Gunji, J., Camargo, M. E., Ceravolo, A., and da Silva, L. C.,** Glomerular lesions in experimental infections of *Schistosoma mansoni* in *Cebus apella* monkeys, *Bull. World Health Organization*, 45, 419, 1971.
122. **Natali, P. G. and Cioli, D.,** Immune complex nephritis in mice infected with *Schistosoma mansoni*, *Fed. Proc.*, 33, 757, 1974.
123. **Hillyer, G. V. and Lewert, R. M.,** Studies on renal pathology in hamsters infected with *Schistosoma mansoni* and *S. japonicum*, *Am. J. Trop. Med. Hyg.*, 23, 404, 1974.
124. **Cavallo, T., Galvanek, E. G., Ward, P. A., and von Lichtenberg, F.,** The nephropathy of experimental hepatosplenic schistosomiasis, *Am. J. Pathol.*, 76, 433, 1974.
125. **Jones, C. E., Rachford, F. W., Ozcel, M. A., and Lewert, R. M.,** *Schistosoma japonicum*: semiquantitative assessment of circulating immune complexes, serum C1q and C3, and their relationship to renal pathology and hepatic fibrosis in rabbits, *Exp. Parasitol.*, 42, 221, 1977.
126. **Culbertson, J. T.,** The cercaricidal action of normal serums, *J. Parasitol.*, 22, 111, 1936.
127. **Standen, O. D.,** The *in vitro* effect of normal and immune serum upon the cercariae of *Schistosoma mansoni*, *J. Helminthol.*, 26, 25, 1952.
128. **Machado, A. J., Gazzinelli, G., Pellegrino, J., and Dias da Silva, W.,** *Schistosoma mansoni:* the role of the complement C3-activating system in the cercaricidal action of normal serum, *Exp. Parasitol.*, 38, 20, 1975.
129. **Greenblatt, H. C., Eveland, L. K., and Morse, S. I.,** *Schistosoma mansoni:* complement and cercarial tail loss during *in vitro* transformation, *Exp. Parasitol.*, 48, 100, 1979.
130. **Kusel, J. R.,** The penetration of human epidermal sheets by the cercariae of *Schistosoma mansoni* and the collection of schistosomula, *Parasitology*, 60, 89, 1970.
131. **Sher, A., Kusel, J. R., Perez, H., and Clegg, J. A.,** Partial isolation of a membrane antigen which induces the formation of antibodies lethal to schistosomes cultured *in vitro*, *Clin. Exp. Immunol.*, 18, 357, 1974.
132. **Tavares, C. A. P., Gazzinelli, G., Mota-Santos, T. A., and Dias da Silva, W.,** *Schistosoma mansoni:* complement-mediated cytotoxic activity *in vitro* and effect of decomplementation on acquired immunity in mice, *Exp. Parasitol.*, 46, 145, 1978.
133. **Santoro, F., Lachmann, P. J., Capron, A., and Capron, M.,** Activation of complement by *Schistosoma mansoni* schistosomula: killing of parasites by the alternative pathway and requirement of IgG for classical pathway activation, *J. Immunol.*, 123, 1551, 1979.
134. **Clegg, J. A. and Smithers, S. R.,** The effects of immune rhesus monkey serum on schistosomula of *Schistosoma mansoni* during activation *in vitro*, *Int. J. Parasitol.*, 2, 79, 1972.
135. **Murrell, K. D. and Clay, B.,** *In vitro* detection of cytotoxic antibodies to *Schistosoma mansoni* schistosomules, *Am. J. Trop. Med. Hyg.*, 21, 569, 1972.
136. **Smith, M. and Webbe, G.,** Damage to schistosomula of *Schistosoma haematobium in vitro* by immune baboon and human sera and absence of cross-reaction with *Schistosoma mansoni*, *Trans. R. Soc. Trop. Med. Hyg.*, 68, 70, 1974.
137. **Kassis, A. I., Warren, K. S., and Mahmoud, A. A. F.,** Antibody-dependent complement-mediated killing of schistosomula in intraperitoneal diffusion chambers in mice, *J. Immunol.*, 123, 1659, 1979.
138. **Gazzinelli, G., Ramalho-Pinto, F. J., and Dias da Silva, W.,** *Schistosoma mansoni:* generation of anaphylatoxin by cercarial extracts, *Exp. Parasitol.*, 26, 86, 1969.
139. **Dean, D. A., Wistar, R., and Murrell, K. D.,** Combined *in vitro* effects of rat antibody and neutrophilic leukocytes on schistosomula of *Schistosoma mansoni*, *Am. J. Trop. Med. Hyg.*, 23, 420, 1974.
140. **Dean, D. A., Wistar, R., and Chen, P.,** Immune response of guinea pigs to *Schistosoma mansoni*. I. *In vitro* effects of antibody and neutrophils, eosinophils and macrophages on schistosomula, *Am. J. Trop. Med. Hyg.*, 24, 74, 1975.

141. **Anwar, A. R. E., Smithers, S. R., and Kay, A. B.**, Killing of schistosomula of *Schistosoma mansoni* coated with antibody and/or complement by human leukocytes *in vitro:* requirement for complement in preferential killing by eosinophils, *J. Immunol.*, 122, 628, 1979.
142. **Sher, A.**, Complement-dependent adherence of mast cells to schistosomula, *Nature (London)*, 263, 334, 1976.
143. **Ottesen, E. A., Stanley, A. M., Gelfand, J. A., Gadek, J. E., Frank, M. M., Nash, T. E., and Cheever, A. W.**, Immunoglobulin and complement receptors on human eosinophils and their role in cellular adherence to schistosomules, *Am. J. Trop. Med. Hyg.*, 26 (Suppl.), 134, 1977.
144. **Ramalho-Pinto, F. J., McLaren, D. J., and Smithers, S. R.**, Complement-mediated killing of schistosomula of *Schistosoma mansoni* by rat eosinophils *in vitro*, *J. Exp. Med.*, 147, 147, 1978.
145. **McLaren, D. J., Ramalho-Pinto, F. J., and Smithers, S. R.**, Ultrastructural evidence for complement and antibody-dependent damage to schistosomula of *Schistosoma mansoni* by rat eosinophils *in vitro*, *Parasitology*, 77, 313, 1978.
146. **McLaren, D. J. and Ramalho-Pinto, F. J.**, Eosinophil-mediated killing of schistosomula of *Schistosoma mansoni in vitro:* synergistic effect of antibody and complement, *J. Immunol.*, 123, 1431, 1979.
147. **Kabil, S. M.**, Host complement in the schistosomal tegument, *J. Trop. Med. Hyg.*, 79, 205, 1976.
148. **Butterworth, A. E.**, Effector mechanisms against schistosomes *in vitro*, *Am. J. Trop. Med. Hyg.*, 26 (Suppl.), 29, 1977.
149. Intestinal nematodes, in *Tropical Medicine*, Hunter, G. W., III, Swartzwelder, J. C., and Clyde, D. F., Eds., W. B. Saunders, Philadelphia, 1976, 454.
150. **Jones, V. E. and Ogilvie, B. M.**, Protective immunity to *Nippostrongylus brasiliensis:* the sequence of events which expels worms from the rat intestine, *Immunology*, 20, 549, 1971.
151. **Ziprin, R. and Jeska, E. L.**, Humoral factors affecting mouse peritoneal cell adherence reactions to *Ascaris suum*, *Infect. Immun.*, 12, 499, 1975.
152. **Leventhal, R. and Soulsby, E. J. L.**, *Ascaris suum:* cuticular binding of the third component of complement by early larval stages, *Exp. Parasitol.*, 41, 423, 1977.
153. **Leventhal, R., Bonner, H., Soulsby, E. J. L., and Schrieber, A. D.**, The role of complement in *Ascaris suum* induced histopathology, *Clin. Exp. Immunol.*, 32, 69, 1978.
154. **Stankiewicz, M. and Jeska, E. L.**, Leucocytes and *Trichinella spiralis*. III. The importance of heat labile and heat stable substances in peritoneal exudate fluid for cell adherence reactions to infective larvae, *Immunology*, 25, 827, 1973.
155. **Stankiewicz, M.**, The participation of $C'3$ and Mg^{++} in the mediation of cell adherence reactions (CAR) to *Trichinella spiralis* in normal non-immune sera, *Bull. Acad. Pol. Sci. (Biol)*, 23, 87, 1975.
156. **Ray, D. and Saha, K.**, Serum immunoglobulin and complement levels in tropical pulmonary eosinophilia, and their correlation with primary and relapsing stages of the illness, *Am. J. Trop. Med. Hyg.*, 27, 503, 1978.

Chapter 7

COMPLEMENT AND INFECTIOUS DISEASES

In 1914 Gunn[1] measured hemolytic complement activity in sera from patients with various bacterial infections and stated that "In all diseases, especially acute infectious diseases, it may be assumed as probable that, in the development of immunity (resistance), at the most active stage of the disease such different types of complement as exist in the patient may be used up by the union of the causal bacteria (antigen) and with their specific immune body." He was only partially successful in demonstrating any particular pattern to complement values during these various infections, though he did note that levels were generally increased at the height of illness and declined to normal with convalescence. Some of his patients had low complement values at various times during infection. Subsequent studies of similarly heterogeneous groups of patients with infections yielded equally heterogeneous results.[2,3] Gunn[1] was probably right; complement activation may well be routine during acute and perhaps even chronic infections. Using a more sensitive assay, Palestine and Klemperer[4] demonstrated factor B activation even in patients with relatively mild sepsis.

Most bacteria and other microorganisms can, under proper experimental conditions, interact with complement through one or both pathways. Tissue injury releases factors capable of activating complement. In one sense, complement activation may be considered a normal reaction to infection and part of the normal inflammatory response. Increased complement synthesis and utilization undoubtedly is of great importance to host defenses. But as has clearly been demonstrated, complement activation may also mediate many of the disadvantageous manifestations of infection. Determination of the exact role or net effect of the complement system in any given patient or in any given disease becomes exceedingly complex.

I. COMPLEMENT AS A NORMAL HOST DEFENSE SYSTEM

An extensive literature attests to the potential of complement as one of the primary defenses against a variety of microorganisms.* Nonvirulent organisms or those of relatively low virulence are more likely to be lysed or opsonized by the complement system, with a minimal requirement for highly specific antibody. These reactions may occur in the absence of antibody via the alternative pathway or even, as in the case of certain viruses, directly through the classical pathway. Virulence is associated with the microbe's development of an ability to evade relatively nonspecific complement-mediated killing. Conversely, immunity represents the host's adaptation to establish primary defenses of a more specific and effective nature.

Examples are numerous. Rough Gram-negative bacteria readily react with complement and are nonvirulent, whereas smooth organisms are less reactive and more virulent; unencapsulated staphylococci and pneumococci behave likewise. Viruses nonpathogenic for man tend to react directly with the complement system, often in the absence of antibody; trypanosomes nonpathogenic for man are readily lysed in normal human serum. Somewhat related are observations that microbial forms such

* For references to examples of interactions between complement and specific microbes, one is referred to the sections on those organisms. References in this chapter will be limited to particularly good examples of a general concept, to studies utilizing a broad range of organisms, or to studies of disease syndromes not limited to one organism.

as schistosomal larvae, which are first reactive with the host defenses, are more likely to be reactive with the complement system.

In situations where antibody plays the central role in host defenses, an enhancing role for complement can often be demonstrated. Complement renders more efficient the interactions of phagocytes and microbes. Complement-dependent roles for basophils and eosinophils have been newly demonstrated in recent years, especially in regard to defenses against parasites such as schistosomes.

The alternative pathway appears to provide a primitive but basic early line of defense against microbial invasion, particularly into the blood stream. However, differing kinetics clearly suggest that, when functional, the classical pathway is far more efficient and more effective than the alternative pathway.[5,6]

In spite of the fact that susceptibility of an organism to complement-mediated microbial lysis favors in vivo clearance,[7-9] even serum-resistant organisms can be effectively cleared by host defenses. Thus, it seems likely that the opsonic functions of antibody and complement are far more important than the lytic functions.[7,8] Such a conclusion is supported by observations of infectious diseases occurring in complement-deficient patients. With the exception of neisserial infections in patients with terminal component deficiencies, only patients who lack C3 function have clearly defective antimicrobial defenses. Thus the key role of complement as an opsonin of normal and immune serum is supported by this review.

However, although most organisms are opsonized by complement in vitro, only a few organisms produce infections in C3 deficiency (pneumococci, streptococci, *Haemophilus influenzae*, proteus, and pseudomonas). Such a consideration stresses the profusion and complexity of the normal immune system. Though complement surely is an important host defense and its interaction with microbes one determinant of virulence, multiple other systems usually compensate satisfactorily for ineffective complement-mediated defense.

II. COMPLEMENT AS A MEDIATOR OF DISEASE

A. Sepsis

Complement activation occurs in a number of syndromes of sepsis: Gram-negative sepsis,[10-13] pneumococcal pneumonia and bacteremia,[14,15] cryptococcal sepsis,[16] and dengue shock syndrome.[17,18] Localized infections, such as empyema, have also been associated with local complement activation.[19] In spite of these observations, it remains to be established that complement activation mediates pathologic sequelae of the infections or, if so, which sequelae. Some pathological role of complement in these complex diseases nevertheless seems likely.

B. Immune-Complex Disease Mediated by Complement

Many infections have secondary immune-complex-mediated phenomena. Nephritis occurs in secondary syphilis (*Treponema pallidum*), *Staphylococcus epidermidis* infection of ventriculoatrial shunts, *Streptococcus pyogenes* pharyngitis or skin infections, pneumococcal pneumonia, corynebacterial and propionibacterial infection of intravascular devices, hepatitis B, cytomegalovirus infection, infectious mononucleosis, malaria, African trypanosomiasis, and schistosomiasis. There is perhaps nothing specific about nephritis with most of these infections; it may be that most infections of any duration at all are associated with immune complexes in the circulation and in the kidneys. Such a notion is supported by studies of renal pathology in patients with typhoid fever[20] and *Schistosoma mansoni* infection[21] with no clinical evidence of kidney disease. Glomeruli, which by light microscopy ranged from normal to clearly nephritic, contained electron-dense granular deposits on or in basement membranes. Immunofluorescence microscopy demonstrated these deposits to contain immunoglobulins and complement.

Further evidence that almost any organism may be associated with immune-complex disease comes from the following observations. Bayer et al.[22] studied 29 patients with endocarditis due to a variety of bacteria and fungi. Ninety-seven percent of their patients had circulating immune complexes; 48% had manifestations of immune-complex disease involving joints, peripheral blood vessels, kidneys, or spleen. Complement levels in 18 of the patients were depressed when compared to levels obtained during convalescence, but only 9 were clearly abnormal. A number of viral infections, especially chronic infections, have been associated with immune-complex disease.[23,24] Beaufils et al.[25] described 11 patients with visceral abscesses ranging from maxillary sinusitis to lung abscess to peritonitis and due to a variety of organisms. At least 6 of these 11 patients had low complement values and all had glomerulitis and acute renal failure of varying degrees. Immunofluorescence microscopy demonstrated glomerular deposits of immunoglobulins and complement. Three papers[26-28] have described the association of immune-complex phenomena (arthritis, intravascular hemolysis, thrombocytopenia and leukopenia, and nephritis) and intestinal-bypass operations (for therapy of morbid obesity). Some of the circulating immune complexes contained antibodies to *Escherichia coli* or *Bacteroides fragilis*.[26] It was assumed that bacterial overgrowth in bowel segments excluded from normal flow led to chronic bacteremia or antigenemia.

In 1916 Longcope[29] reviewed the subject of serum sickness. At that time, treatment of a number of diseases included transfusion of antiserum, usually raised in horses. The syndrome developed 6 or more days after injection of the foreign antiserum. Lymph node swelling was followed by a pruritic skin rash, edema, fever, malaise, and headache. Arthralgias and splenomegaly were common. There were "often disturbances in the functional activities of the kidneys" and proteinuria was noted in 5 to 10% of cases. Longcope[29] suggested that a similar phenomenon might underlie numerous medical illnesses including allergies, and he suggested that "allergies" may explain some manifestations of infectious diseases. Specifically he mentioned cutaneous hemorrhages, subcutaneous nodules, nephritis, and arthritis associated with chronic streptococcal infections.

More recently, Cochrane has reviewed mechanisms of immune-complex deposition and has stressed the role of complement.[30,31] Size of immune complexes is critical in determining whether they precipitate and thus can be ingested by phagocytes or remain soluble to be filtered and deposited in blood vessel walls, especially those of the renal glomerulus. Size may depend upon variables of the antigen and also on the nature of the host response, which includes classes and amounts of immunoglobulins produced. In patients with hepatitis B and chronic active hepatitis,[32,33] immune-complex-mediated arthritis was most clearly related to concentrations of immunoglobulins in the complexes (lower concentrations were associated with lesser degrees of tissue injury), with the specific immunoglobulin classes and subclasses in the complexes (the presence of IgA and IgM was especially toxic; subclasses of IgG that did not activate complement were less likely to produce damage) and with the presence or absence of complement in the complexes. Complement and complement-mediated neutrophil responses are critical to the tissue damage induced by immune complexes.[30,31]

A number of investigators have studied the hypothesis that streptococcal disease may be especially prone to induce a serum sickness-like illness.[29] Kellett[34] demonstrated low hemolytic complement levels in patients with acute glomerulonephritis. (Data were not available to establish the etiologic role of streptococci.) Patients with chronic nephritis did not have low complement levels.[34,35] Reader[36] obtained similar results; sequential studies indicated that complement levels were initially low and returned to normal within 20 days of onset of nephritis. Complement levels did not correlate with severity of illness. Lange et al.[37] reported similar results in patients with acute nephritis and the nephrotic syndrome. The immunopathology and immunopathogenesis of poststreptococcal nephritis remain a subject of some controversy,[38] but a role for immune complexes and complement seems secure. That the nature of the antigen is critical in the induction of immune-complex nephritis is implicit in the particular association of streptococcal infection and *clinically important* nephritis. Though, as noted above, many if not all infections may induce pathological changes of nephritis, few regularly produce such prominent disease as streptococci do. Similarly, Takekoshi et al.[39] have suggested a consistent association between a specific kind of nephritis, membranous nephropathy, and chronic viremia with a specific virus, hepatitis B. Finally, O'Connor et al.[40] have emphasized the curious fact that, whereas infection-associated immune-complex phenomena occur commonly with chronic endocarditis, *Staphylococcus aureus* infection is unusual among *acute* endocarditides in the frequency of immune-complex disorders. Of 14 patients with *S. aureus* endocarditis in their study, 11 had nephritis (30% of the patients infected with other organisms had nephritis). Furthermore, nephritis correlated with hypocomplementemia and thrombocytopenia. The characteristics of specific microbes that make them more likely to induce immune-complex disease are not well understood.

C. Other Mechanisms of Complement-Mediated Inflammation

Oldstone and Dixon[41] have emphasized that complement-mediated lysis of virus-infected cells may be one mechanism of host injury during viral infections. This phenomenon has been most clearly demonstrated with lymphocytic choriomeningitis virus infection.

Several allergic lung diseases may be due to a complement-mediated inflammatory reaction, analogous to the Arthus reaction. Pulmonary symptoms are thought to follow inhalation of infectious agents by patients who have circulating antibodies to those agents or in whom agents interact with the alternative pathway. Organisms include bacteria, such as lactobacilli[42] and Micromonosporaceae,[43] and fungi, such as aspergillus and penicillium species.[43,44]

Complement activation by generally nonpathogenic or mildly pathogenic propionibacteria may mediate the inflammatory events of acne vulgaris.[45,46] Contributions of complement chemotactic factors and resultant neutrophil accumulation have been stressed.

Interest has been directed to the potential role of complement, complement chemotactic factors, and polymorphonuclear leukocytes in inflammatory periodontal disease.[47-50] It is thought that antibodies and complement in crevices of the gingival tissues[47] interact with microorganisms common to this area, especially *Streptococcus mutans*. Resultant ingress of leukocytes would, by release of lysomal enzymes, damage the periodontal tissue. It has also been suggested that bacteria may, by interaction with the alternative pathway, produce this phenomenon independently of antibody.[48] Other authors have however suggested that complement may not play a critical role.[49]

III. COMMENTS

A challenge for investigators is the more complete elucidation of competing contributions by complement to host defenses and inflammatory host damage in infections associated with complement activation. If manipulation of the complement system becomes a clinical reality, such information will be critical.

REFERENCES

1. **Gunn, W. C.**, The variation in the amount of complement in the blood in some acute infectious diseases and its relation to the clinical features, *J. Pathol. Bacteriol.*, 19, 155, 1914.
2. **Hadjopoulos, L. G. and Burbank, R.**, The role of complement in health and disease. A clinical study of the hemolytic complement of human sera, *J. Lab. Clin. Med.*, 14, 131, 1928–1929.
3. **Ecker, E. E., Seifter, S., Dozois, T. F., and Barr, L.**, Complement in infectious disease in man, *J. Clin. Invest.*, 25, 800, 1946.
4. **Palestine, A. G. and Klemperer, M. R.**, *In vivo* activation of properdin Factor B in normotensive bacteremic individuals, *J. Immunol.*, 117, 703, 1976.
5. **Traub, W. H. and Kleber, I.**, Selective activation of classical and alternative pathways of human complement by "promptly serum-sensitive" and "delayed serum-sensitive" strains of *Serratia marcescens*, *Infect. Immun.*, 13, 1343, 1976.
6. **Giebink, G. S., Verhoef, J., Peterson, P. K., and Quie, P. G.**, Opsonic requirements for phagocytosis of *Streptococcus pneumoniae* types VI, XVIII, XXIII, and XXV, *Infect. Immun.*, 18, 291, 1977.
7. **Roantree, R. J. and Rantz, L. A.**, A study of the relationship of the normal bactericidal activity of human serum to bacterial infection, *J. Clin. Invest.*, 39, 72, 1960.
8. **Roantree, R. J. and Pappas, N. C.**, The survival of strains of enteric bacilli in the blood stream as related to their sensitivity to the bactericidal effect of serum, *J. Clin. Invest.*, 39, 82, 1960.
9. **Simberkoff, M. S., Ricupero, I., and Rahal, J. J., Jr.**, Host resistance to *Serratia marcescens* infection: serum bactericidal activity and phagocytosis by normal blood leukocytes, *J. Lab. Clin. Med.* 87, 206, 1976.
10. **McCabe, W. R.**, Serum complement levels in bacteremia due to Gram-negative organisms, *N. Engl. J. Med.*, 288, 21, 1973.
11. **Fearon, D. T., Ruddy, S., Schur, P. H., and McCabe, W. R.**, Activation of the properdin pathway of complement in patients with Gram-negative bacteremia, *N. Engl. J. Med.*, 292, 937, 1975.
12. **Füst, G., Petras, G., and Ujhelyi, E.**, Activation of the complement system during infections due to Gram-negative baceteria, *Clin. Immunol. Immunopathol.*, 5, 293, 1976.
13. **Bjornson, A. B., Altemeier, W. A., and Bjornson, H. S.**, The septic burned patient. A model for studying the role of complement and immunoglobulins in opsonization of opportunist micro-organisms, *Ann. Surg.*, 189, 515, 1979.
14. **Reed, W. P., Davidson, M. S., and Williams, R. C., Jr.**, Complement system in pneumococcal infections, *Infect. Immun.*, 13, 1120, 1976.
15. **Coonrod, J. D. and Rylko-Bauer, B.**, Complement levels in pneumococcal pneumonia, *Infect. Immun.*, 18, 14, 1977.
16. **Macher, A. M., Bennett, J. E., Gadek, J. E., and Frank, M. M.**, Complement depletion in cryptococcal sepsis, *J. Immunol.*, 120, 1686, 1978.
17. **Russell, P. K., Intavivat, A., and Kanchanapilant, S.**, Antidengue immunoglobulins and serum $\beta 1$ c/a globulin levels in dengue shock syndrome, *J. Immunol.*, 102, 412, 1969.
18. **Bokisch, V. A., Top, F. H., Jr., Russell, P. K., Dixon, F. J., and Müller-Eberhard, H. J.**, The potential pathogenic role of complement in dengue hemorrhagic shock syndrome, *N. Engl. J. Med.*, 289, 996, 1973.
19. **Lew, P. D., Zubler, R., Vaudaux, P., Farquet, J. J., Waldvogel, F. A., and Lambert, P.-H.**, Decreased heat-labile opsonic activity and complement levels associated with evidence of C3 breakdown products in infected pleural effusions, *J. Clin. Invest.*, 63, 326, 1979.

20. Sitprija, V., Pipatanagul, V., Boonpucknavig, V., and Boonpucknavig, S., Glomerulitis in typhoid fever, *Ann. Intern. Med.*, 81, 210, 1974.
21. de Silva, L. C., de Brito, T., Camargo, M. E., de Boni, D. R., Lopes, J. D., and Gunji, J., Kidney biopsy in the hepatosplenic form of infection with *Schistosoma mansoni* in man, *Bull. W.H.O.*, 42, 907, 1970.
22. Bayer, A. S., Theofilopoulos, A. N., Eisenberg, R., Dixon, F. J., and Guze, L. B., Circulating immune complexes in infective endocarditis, *N. Engl. J. Med.*, 295, 1500, 1976.
23. Oldstone, M. B. A. and Dixon, F. J., Immune complex disease in chronic viral infections, *J. Exp. Med.*, 134, 32S, 1971.
24. Oldstone, M. B. A., Virus neutralization and virus-induced immune complex disease. Virus-antibody union resulting in immunoprotection or immunologic injury—two sides of the same coin, *Prog. Med. Virol.*, 19, 84, 1975.
25. Beaufils, M., Morel-Maroger, L., Sraer, J.-D., Kaufer, A., Kourilsky, O., and Richet, G., Acute renal failure of glomerular origin during visceral abscesses, *N. Engl. J. Med.*, 295, 185, 1976.
26. Wands, J. R., LaMont, J. T., Mann, E., and Isselbacher, K. J., Arthritis associated with intestinal-bypass procedure for morbid obesity. Complement activation and characterization of circulating cryoproteins, *N. Engl. J. Med.*, 294, 121, 1976.
27. Moake, J. L., Kageler, W. V., Cimo, P. L., Blakely, R. W., Rossen, R. D., and Haesse, W., Intravascular hemolysis, thrombocytopenia, leukopenia, and circulating immune complexes after jejunal-ileal bypass surgery, *Ann. Intern. Med.*, 86, 576, 1977.
28. Drenick, E. J., Stanley, T. M., Border, W. A., Zawada, E. T., Dornfeld, L. P., Upham, T., and Llach, F., Renal damage with intestinal bypass, *Ann. Intern. Med.*, 89, 594, 1978.
29. Longcope, W. T., The susceptibility of man to foreign proteins, *Am. J. Med. Sci.*, 152, 625, 1916.
30. Cochrane, C. G., The role of immune complexes and complement in tissue injury, *J. Allerg.*, 42, 113, 1968.
31. Cochrane, C. G., Mechanisms involved in the deposition of immune complexes in tissues, *J. Exp. Med.*, 134, 75S, 1971.
32. Wands, J. R., Mann, E., Alpert, E., and Isselbacher, K. J., The pathogenesis of arthritis associated with acute hepatitis-B surface antigen-positive hepatitis. Complement activation and characterization of circulating immune complexes, *J. Clin. Invest.*, 55, 930, 1975.
33. Wands, J. R., Alpert, E., and Isselbacher, K. J., Arthritis associated with chronic active hepatitis: complement activation and characterization of circulating immune complexes, *Gastroenterology*, 69, 1286, 1975.
34. Kellett, C. E., Complement titre in acute nephritis. With special reference to causation by reversed anaphylaxis, *Lancet*, 2, 1262, 1936.
35. Kellett, C. E. and Thomson, J. G., Complementary activity of blood serum in nephritis, *J. Pathol. Bacteriol.*, 48, 519, 1939.
36. Reader, R., Serum complement in acute nephritis, *Br. J. Exp. Pathol.*, 29, 255, 1948.
37. Lange, K., Graig, F., Oberman, J., Slobody, L., Ogur, G., and LoCasto, F., Changes in serum complement during the course and treatment of glomerulonephritis, *Arch. Intern. Med.*, 88, 433, 1951.
38. Nissenson, A. R., Baraff, L. J., Fine, R. N., and Knutson, D. W., Poststreptococcal acute glomerulonephritis: fact and controversy, *Ann. Intern. Med.*, 91, 76, 1979.
39. Takekoshi, Y., Tanaka, M., Shida, M., Satake, Y., Saheki, Y., and Matsumoto, S., Strong association between membranous nephropathy and hepatitis-B surface antigenaemia in Japanese children, *Lancet*, 2, 1065, 1978.
40. O'Connor, D. T., Weisman, M. H., and Fierer, J., Activation of the alternate complement pathway in *Staph. aureus* infective endocarditis and its relationship to thrombocytopenia, coagulation abnormalities, and acute glomerulonephritis, *Clin. Exp. Immunol.*, 34, 179, 1978.
41. Oldstone, M. B. A. and Dixon, F. J., Acute viral infection: tissue injury mediated by anti-viral antibody through a complement effector system, *J. Immunol.*, 107, 1274, 1971.
42. Huis in 't Veld, J. H. J. and Berrens, L., Inactivation of hemolytic complement in human serum by an acylated polysaccharide from a Gram-positive rod: possible significance in pigeon-breeder's disease, *Infect. Immun.*, 13, 1619, 1976.
43. Marx, J. J., Jr. and Flaherty, D. K., Activation of the complement sequence by extracts of bacteria and fungi associated with hypersensitivity pneumonitis, *J. Allerg. Clin. Immunol.*, 57, 328, 1976.
44. Olenchock, S. A. and Burrell, R., The role of precipitins and complement activation in the etiology of allergic lung disease, *J. Allerg. Clin. Immunol.*, 58, 76, 1976.
45. Webster, G. F., Leyden, J. J., Norman, M. E., and Nilsson, U. R., Complement activation in acne vulgaris: *in vitro* studies with *Propionibacterium acnes* and *Propionibacterium granulosum*, *Infect. Immun.*, 22, 523, 1978.

46. **Webster, G. F., Leyden, J. J., and Nilsson, U. R.**, Complement activation in acne vulgaris: consumption of complement by comedones, *Infect. Immun.*, 26, 183, 1979.
47. **Shillitoe, E. J. and Lehner, T.**, Immunoglobulins and complement in crevicular fluid, serum and saliva in man, *Arch. Oral Biol.*, 17, 241, 1972.
48. **Allison, A. C., Schorlemmer, H. U., and Bitter-Suermann, D.**, Activation of complement by the alternative pathway as a factor in the pathogenesis of periodontal disease, *Lancet*, 2, 1001, 1976.
49. **McArthur, W. P., Baehni, P., and Taichman, N. S.**, Interactions of inflammatory cells and oral microorganisms. III. Modulation of rabbit polymorphonuclear leukocyte hydrolase release response to *Actinomyces viscosus* and *Streptococcus mutans* by immunoglobulins and complement, *Infect. Immun.*, 14, 1315, 1976.
50. **Patters, M. R., Schenkein, H. A., and Weinstein, A.**, A method for detection of complement cleavage in gingival fluid, *J. Dent. Res.*, 58, 1620, 1979.

INDEX

A

Abscesses, amebic, 124
Acholeplasma laidlawii, 70—71
Acne, propionibacteria and, 66
Acne vulgaris, 144
Actinomycetes, and farmers' lung, 67—68
Acute phase reaction, 21
Agammaglobulinemai, 12
AKR mouse leukemia virus, 105
Allergic lung diseases, 144
Allergic reactions
    immune complexes in, 143
    parasites and, 115
Alphaviruses, 102
Alternative pathway, 4—6
    allergic lung disease, 144
    assessment of, 21
    bacterial surfaces and, 46
    deficiencies, 32
    hemolytic assays, 20
    immune adherence, 14
    intracellular killing, 13
    lysis, 18
    opsonization, 12
    P. aeruginosa and, 47
    sickle cell disease, 37
    staphylococcal encapsulation and, 57
    Staphylococcus aureus, 56
Amebiasis, 116
Amebic colitis, 124
Anaphylactoid purpura, classical pathway deficiencies, 31
Anaphylatoxins, 16, 72
    C3a as, 17
    C5a as, 6, 17
Anaplasma marginale, 118
Ancylostoma species, 117
Anemia, African, trypanosamiasis and, 120
Angiostrongylus cantonensis, 117
Anisakis marina, 117
Antibodies, see also Immune complexes; Immunoglobulin(s)
    alternative pathway activation, 5—6
    fluorescein tagged, in assays, 20
    19S, and *Shigellae*, 51
Antibody-dependent cell-mediated cytotoxicity (ADCC), 15, 101
Antigen-antibody-C142 complexes, 18
Antigens, see also Immune complexes
    capsular, see also K antigens, 45—46
    E. coli, 49—50
    Gram-negative bacteria, 45
    helminth, 125
    hepatitis B, 107—108
    influenza viruses, 104
    variant
        of protozoa, 115
        trypanosomes, 120, 121
    viral infection and, 95, 96
    viral surface, 96
Arboviruses, 102
Arenaviruses, 106—107
Argentine hemorrhagic fever, 107
Arginine, anaphylatoxin, 17
Arthritis
    hepatitis B and, 108, 143
    immune complexes and, 142, 143
Arthus reaction, 16, 67, 144
Ascariasis, 117
Ascaris suum, 133
Aspergillus, 85, 86, 144
Assays, complement, 20—21

B

Babesiosis, 116, 119
Bacillus proteus, 12
Bacteremia
    E. coli, 50
    pneumococcal, and complement activation, 142
Bacteria, see also specific bacteria
    chemotaxis, 17
    classification, 43, 44
    lysis of, 18—19
    surfaces, 43—46
Bacteroides, 53
Bacterioides fragilis, 142
Bacteriophages, 31, 108—109
Balantidium coli, 116
Basophils, see Mast cells
Beta hemolytic streptococci, see also streptococcal species
    C3 deficiency and, 32
    infections, 60—61
Blastomyces dermatidis, 86
B-lymphocyte
    C3 receptors, 14—15
    C3b receptors of, 3
    chemotactic factors, 17
    Epstein-Barr virus and, 100
Bovine rhinotracheitis virus, infectious, 97
Burkitt's lymphoma, 100

C

C1
    erythrocyte phagocytosis, 10
    in alternative pathway, 5
    in classical pathway, 4
C1 esterase inhibitor, 17
C1q, and opsonization, 12
C2
    alternative pathway, 5
    classical pathway, 4
    deficiency, 31—32

erythrocyte phagocytosis, 10
heat inactivation of, 22
C3, 1—3, 4
   alternative pathway, 5
   deficiency, 32
   endotoxin, 73, 74
   erythrocyte phagocytosis, 10
   hypercatabolism, 32
   nephritic factor, 33
   pneumococcal pneumoniae and, 63
   protein-calorie malnutrition and, 38
   receptors for, and Epstein Barr virus, 100
   schistosomes and, 132
   splenectomy and, 38
C3
   staphylococcal infections, 58, 60
   Staphylococcus aureus and, 56—57
   virus neutralization, 96
C3a, and chemotaxis, 16
C3 convertase, 3
C3b
   and C5, 6
   and chemotaxis, 17
   receptor interaction, 15
C3b inactivator, hereditary absence of, 32
C3e, and leukocyte mobilization, 17
C4
   alternative pathway, 5
   C3 receptor, 15
   classical pathway, 4
   erythrocyte phagocytosis, 10
   opsonization, 12
C5
   C3b and, 6
   deficiency of, 33
   dysfunction, 35—36
   intracellular killing, 13
   pneumococcal pneumoniae and, 63
   staphylococcal infection, 60
C5a, and chemotaxis, 16
C5b, in terminal sequence, 6
C5—9, and erythrocyte lysis, 18
C6
   endotoxin, 73
   deficiency, 34
   terminal sequence, 6
C7
   deficiency, 34
   terminal sequence, 6
C8
   deficiency, 34
   terminal sequence, 6
Cancer therapy, with *Corynebacterium parvum*, 66
Candida, 86—87
   C5 deficiency and, 33
   complement activation, 89
Capillariasis, 117
Capsoid, immunological interaction with, 96
Capsule, 45—46
   antigens, see K antigens
   cryptococcal, 85, 87—88

Pneumococcus pneumoniae, 64—65
staphylococcal opsonization, 57—58, 59
Carboxypeptidase B, 16
Catabolism, complement, 1
Cell membrane, helminth, 125
Cells, lysis of, 18, 101
Cell walls, see also Teichoic acids; Membranes, 43
   and lysis, 19—20
   streptococcal groups, 61—62
Cercariae, schistosomal, 129—131
Cestodes, 116, 126—127
CH50 assay, 20
Chagas' disease, 116, 122—123
Chemiluminescence, phagocytosis assay, 10
Chemotaxis
   candida species, 87
   complement components in, 16—17
   staphylococcal peptidoglycans and, 60
Chitin, fungal, 85
Chlicero ulcer, 116
Chronic active hepatitis, 108, 142
Classical pathways, 4
   assessment of, 21
   complement activation, 12
   deficiencies of, 31
   lysis, 18
   deficiences of, 31
   Staphylococcus aureus and, 56—57
Coagulation, complement and, 18
Coagulopathies, see also Disseminated intravascular coagulopathy
   arenaviruses and, 106—107
   endotoxins and, 72
Cobra venom factor, and complement inactivation, 21—22
Coccidioides immitis, 86, 89
Coccidiosis, 116
Colitis, amebic, 124
Complement
   assays, 20—21
   assessment of pathways, 21
   coagulation, 18
   heat lability of, 22
   inactivation of, 22
   in infectious diseases, strategies of study, 21
   mediator of disease, 142—144
   methodology, 20
   normal host defense system, 141—142
   synthesis, 1
Complement activation
   cryptococci and, 88—89
   endotoxin and, 71—72, 72—73
   fungi and, 89
   Gram-negative bacteria and, see also specific bacteria, 56
   Gram-positive bacteria and, see also specific bacteria, 56
   pneumococcal cell walls, and, 64—65
   pneumococcal pneumonia and, 63—64
   pneumococci, 65
   schistosomiasis and, 130—131
   sepsis syndromes and, 142

staphylococcal encapsulation and, 57—59
Staphylococcus aureus and, 56—57
streptococcal groups and, 61—62
Complement deficiencies
  acquired, 36
  alternative pathway, 32
  C3, 32—33
  C5 dysfunction, 35—36
  classical pathways, 31—32
  terminal component, 33—35
  treatment, 36
Complement-mediated cell lysis, 15, 18, 101
Complement system
  alternative pathway, 2, 4—6
  C3, 1—4
  classical pathway, 24
  terminal sequence, 26
  terminology, 1
Concomitant immunity, schistosomes and, 128—129
Coronaviruses, 107
Corynebacterial infections, 66, 142
Cornyebacterium parvum, 56
Cryoprecipitates, hepatitis B and, 108
Cryptococcal sepsis, 142
Cryptococci
  capsules, 87—88
  complement activation and opsonization, 88—89, 142
Cryptococcus neoformans, 86, 87—89
Cuticle, helminth, 125
Cytochalasin B, and phagocytosis, 11
Cytomegalovirus, 99—100, 142
Cytotoxicity, antibody-dependent cellular, 15, 18, 101

D

Delayed hypersensitivity, 16
Dengue virus, 102—103, 142
Dental plaque, 53
Diplococcus pneumoniae, see *Streptococcus pneumoniae*
Disseminated intravascular coagulation, see also Coagulopathies
  arenaviruses and, 107
  dengue virus and, 103
  endotoxin and, 72
  malaria and, 118

E

Echinococcus, 126—127
ECHO virus, 101
Empyema, and complement activation, 142
Encapsulation, see Capsule
Encephalitis
  herpes simplex virus, 98
  Sindbis virus, 102

Endocarditis
  Escherichia coli, 49
  immune complex, 143
  Staphylococcus aureus, 144
  streptococcal, 60—61, 62
Endotoxemia, 47, 71—74
Endotoxins, 71—74
  complement activation, 71—72, 72—73
  composition of, 71
  Schwartzmann reaction, 73—74
Entamoeba histolytica, 116, 124
Enterobius vermicularis, 117
Enterococcal septicemia, 33
Enteroviruses, 101
Envelope, viral, 93, 96
Eosinophilia
  parasites and, 115
  schistosomula, 131—132
  tropical pulmonary, 133
Epstein-Barr virus, 100
Equine arteritis virus, 97, 103
Equine encephalitis virus, Estern, 102
Erysipelothrix rhusopathiae, 56, 66
Erythema nodosum leprosorum, 67
Erythrocytes
  C3 receptors of, 14
  C3b receptors of, 3
  immune adherence, 14
  lysis of, 18—20
  phagocytosis of, 10—12
  Treponema pallidum adherence to, 14
Escherichia coli
  antibodies to, after intestinal bypass, 142
  encapsulated, 58
  intracellular killing of, 13
  lysis of, 19
  opsonization of, 12
  O side chains, 45
Espundia, 116
Esterase, Cl, 4
Ethyleneglycoltetraacetic acid (EGTA), 12, 22

F

Factor B
  alternative pathway, 5
  deficiency, 31
  farmer's lung, 68
  heat inactivation of, 22
  Pseudomones aeruginosa opsonization, 47
  pneumococcal pneumonia and, 63
  sepsis and, 141
Familial disorders, herpes simplex susceptibility, 99
Farmer's lung, 67—68
Fc, complement binding, see also Receptors, 4
Feline lymphoma virus, 105
Fibrinogen, complement and, 18
Filariasis, 117
Flaviviruses, 102—103
Fluorescein-tagged antibodies, 20

Forssman antigen, 4
Forssman antiserum, 20
Flukes, 117, 127—132
Francisella tularensis, 48
Fungi
    allergic lung disease, 144
    aspergillus, 85
    candida, 86—87
    Coccidioides immitis, 89
    Cryptococcus neoformans, 87—89
    deep, associated with human diseases, 86
    Paracoccidioides brasiliensis, 89
    Torulopsis glabrata, 87
Fusobacteria, 53—54

G

Genome
    cellular, viruses and, 93, 95
    viral, 93
Giardia lamblia, 116
Globulin, 1H, 5
Glomeruli
    C2 receptors of, 15
    C3b receptors of, 3
    trypanosomal immune complexes in, 120
    viral immune complexes in, 96—97
Glomerulonephritis
    corynebacterial, 66
    cytomegalovirus, 100
    ECHO virus and, 101
    Epstein-Barr virus and, 100
    hepatitis B and, 108
    immune complex, 141—142
    intestinal bypass and, 142
    Listeria monocytogenes, 66
    malaria and, 118—119
    pneumococcal infections and, 66
    Propionibacterium acnes, 66
    schistosomiasis and, 128
    Staphylococcus aureus, 144
    Staphylococcus epidermidis, 55—56
    streptococcal, 62
    Streptococcus pyogenes and, 143, 144
    syphilis and, 46—47
Glycocalyx, helminths, 125
Glycoproteins, see also Antigens
    helminth, 125
    Trypanosoma brucei, 120
Gnathostoma species, 117
Gonoccal septicemia, C6 deficiency and, 34
Gonorrhea, disseminated, see also *Neisseria gonorrhoeae*, 54
Gram negative bacteria
    aerobic
        Neisseria gonorrhoeae, 54—55
        Neisseria meningitides, 55
    aerobic rods, 47—48
    anaerobic
        bacteriodes, 53
        fusobacteria, 53—54
    classification of, 44
    facultatively anaerobic rods, 48—53
    lysis of, 18, 19
    serum sensitivity of, 141
Gram-negative sepsis
    and complement activation, 142
    endotoxins and, 71
Gram positive bacteria
    asporogenous rods, 66
    cell walls, 43
    cocci
        Staphylococcus aureus, 56—60
        Staphylococcus epidermidis, 55—56
    streptococci, 60—63
        Streptococcus pneumoniae, 63—66
    resistance to lysis, 18
Gram stain, 43
Granulocyte adherence, C5a and, 17
Granuloucytes, see Polymorphonuclear leucocytes

H

β1H globulin, 5
Haemophilus influenzae, 52—53
    C3 deficiency and, 32
    sickle cell disease, 36
Helminths, 116, 125—133
    cestodes, 126—127
    nematodies, 132—133
    sites of infections with, 125
    trematodes, 127—132
Hemolysis, immune complexes and, 142, 143
Hemolytic assay, 20
Hemolytic streptococcal infections, 32, 60—61
Hemorrhagic fever syndromes, arenaviruses and, 106—107
Hepatitis A virus, 101—102
Heptatits B
    chronic acitive, 108, 142
    immune complex disease, 142—144
Hepatitis B virus, 107—108
Herpes simplex virus, 97—99
Herpes viruses, 98—101
    cytomegalovirus, 99—100
    Epstein-Barr, 100
    herpes simplex, 97—99
    infectious bovine rhinotracheitis, 101
Herpes zoster, C5 deficiency and, 33
Hexose monophosphate shunt, phagocytosis assay, 10
Histamine
    C3a and, 17
    endotoxin and, 71
Histoplasma capsulatum, 86
HLA genes, 31, 34
Hookworm, 117
Hyatid disease, 116, 127
Hydrazine, and complement inactivation, 22
Hymenolepsis, 126
Hypersensitivity lung disease, 85
Hypocomplementemie, and endotoxic shock, 71—73

I

Immune adherence, 14, 15
Immune complexes
  complement mediation in, 142—144
  Epstein-Barr virus and, 100
  mechanics of deposition, 143
  schistosomiasis, 128
  streptococcal glomerulonephritis, 62
  virus and antibody, 96
Immune response
  parasite evasion of, 115
  viral infections, 96—97
Immunochemical assays, 20
Immunoconglutinin
  African trypanosomiasis and, 120
  plasmodia and, 118
Immunoelectrophore, 21
Immunofluorescence, C3 deposition, 21
Immunoglobulin(s)
  classical pathway, 4
  complement-coagulation interaction, 18
  immune complex mediated arthritis, 143
  intracellular killing, 13
  lysis, 18—19
  opsonization, 11—12
  pneumococci and, 66
  S. aureus opsonization, 56
  viruses, 96
Immunoglobulin E, parasites and, 115
Immunoglobulin G
  B lymphocyte receptors, 14—15
  C3 nephritic factor, 33
  lysis, 19
  P. aeruginosa opsonization, 47
  phagocytosis, 11—12
Immunoglobulin M
  and lysis, 19
  and macrophage phagocytosis, 10
Inactivator, C3b, absence of, 32
Infectious mononucleosis, 100, 142
Influenza A virus, 97, 103, 104
Influenza B virus, 103
Inhibitors
  alternative pathway, 5
  C3, 3
  terminology, 1, 2
Intestinal bypass operations, and immune complex disorders, 143
Intrapagocytotic killing, 13—14
Intravascular devices, see Ventriculoatrial shunt
Iospora belli, 116

J

Jarisch-Hexheimer reaction, 47
Junin virus, 107

K

Kala-azar, 116, 124
K antigens, 45—46, 49—50
Killer cells, 101
Klebsiella pneumoniae, 51, 58

L

Lactobacilli, and allergic lung disease, 144
Lancefield classification, of streptococci, 60, 61
Leiner's disease, C5 dysfunction and, 35—36
Leishmania, 100—101, 116
Leprosy lepromatous, 67
Leukemia, and acquired complement deficiencies, 36
Leukemia viruses, 105
Leukocytes, and schistosomes, see also Chemotaxis; Monocytes; Polymorphonuclear leucocytes, 131—132
Leukopenia
  in *Escherichia coli* bacteremia, 50
  intestinal bypass and, 142
Lipid, A, 45, 71
Lipodystrophy partial
  C3 deficiency and, 33
  measles virus and, 104
Lipopolysaccharides
  bacterial, 71—74
  bacteroides, 53
  Fusobacterium nucleatum, 54
  Gram-negative bacteria, 45
  host defenses and, 46
Listeria monocytogenes, 56, 66
L-phase variants, staphylococcal, 58
Lung diseases, allergic, 144
Lymphocytic choriomeningitis virus, 97, 106, 144
Lymphokine, chemotactic, 17
Lymphoma virus, feline, 105
beta-Lysins, 19
Lysis, 18
Lysosomes, virus infection and, 95
Lysozyme, and lysis, 19

M

Macrophages, see also Opsonization, 8, 15
  and chemotaxis, 17
  C3b receptors of, 3
  erythrocyte phagocytosis, 10—11
Malaria, 116—119, 142
Mast cells
  C3a and, 17
  schistosomula, 131
Measles virus, 97, 104—105
Membranes, see also Cell walls; Teichoic acids, 43
  attack mechanism, 6
  lysis, 19—20
Membranous nephropathy hepatitis B and, 108, 144

Meningitis
  cryptococcal, 88
  Haemophilus influenzae, 52—53
  lymphocytic choriomeningitis virus and, 97, 106, 144
  meningococcal, 32—55
Meningoencephalitis, amebic, 116
Metabolic studies, 20—21
Metazoa, see Helminths
Methodology, complement, 20
Micromonosporaceae, and allergic lung disease, 144
Microsporaceae, 67—68
Moloney leukemia virus, 105
Moniliasis, 33, 116
Monocytes, 8
  C3b receptors of, 3
  complement independent phagocytosis by, 10
  intracellular killing by, 13
Mononucleosis, infectious, 100
M protein, streptococcal, 61, 62
Mucocutaneous candidiasis, 86
Mucor species, 86
Multiple sclerosis, measles virus and, 104
Mycobacteria, 67
Mycoplasma canis, 70
Mycoplasmacidal activity, 69
Mycoplasma gallisepticum, 70
Mycoplasma hominis, 70
Mycoplasma pneumoniae, 69—70
Mycoplasmas, classical pathway deficiencies, 31
Mycoses, see Fungi; specific organisms

N

Naegleria species, 116
Necator americanus, 117
Neisseria gonorrhoeae, 34, 54—55
Neisseria meningitidis, 32—35, 55
Nematodes, 117, 132—133
Neutralization, of virus, complement mediated, 96—97
Neutrophils, 8
Nephritic factor, C3 and, 33
Nephritis, see Glomerulonephritis
Nephrotic syndrome, complement function and, 36
Newcastle disease virus, 97, 104
Nippostrongylus brasilienisis, 132—133
Nitroblue tetrazolium, phagocytosis assay, 10
North American blastomycosis, 86
Nucleic acid, viral, 93
Nucleocapsid, viral, 93

O

O antigens, 45, 49
Oesophagostoma species, 117
Oncornaviruses, 105—106

Opsonins, and intracellular killing, 13
Opsonization
  C3-deficiency and, 33
  C3 receptors and, 6—13
  C5 dysfunction and, 36
  cryptococci, 88—89
  Gram-negative bacteria, see also specific bacteria, 56
  Gram-positive bacteria, see also specific bacteria, 56
  pneumococci, 65
  sickle cell disease and, 37
  staphylococcal encapsulation and, 57—59
  Staphylococcus aureus, 56—57
  streptococci, 62, 64, 65
Orthomyxoviruses, 103—104
Otitis media, 31, 32

P

Papovaviruses, 101
Paracoccidioides brasiliensis, 86, 89
Paramyxoviruses, 104—105
Parasites
  diseases of man, 116—117
  evasion of immune responses, 115
  helminths, 125—133
  protozoa, 115—125
  Pasturella tularensis, 48
  toxoplasma, 124—125
Pathogenicity
  capsule and, 45
  O side chains and, 45
Penicillin sensitivity, *Neisseria gonorrhoeae*, 54—55
Penicillium, and allergic lung disease, 144
Peptidoglycans, 43
  host defenses and, 46
  staphylococcal, and complement activation, 58—59
Periodontal disease, 53, 144
Phagocytes, see also Macrophages; Opsonization; Polymorphonuclear leukocytes, 8
Phagocytosis, mechanisms of, see also Opsonization, 9—13
Phycomycoses, 86
Picorna viruses, 101—102
Pinworm infection, 117
Plasmodia, 116—119
Plasmodium berghei, in mice, 117
Plasmodium coatneyi, in rhesus monkeys, 118
Plasmodium cynomolgi, 119
Plasmodium falciparum, in children, 118
Plasmodium gallinaceum, 117
Plasmodium knowlesi
  in monkeys, 117
  serum sensitivity of, 116
Plasmodium malariae, and nephrosis, 119
Platelets
  beta-lysine of, 19
  C3b receptors of, 3, 18

intestinal bypass and, 142
   rabbit, endotoxin and, 72
Pleuropneumonia-like organisms, 69—71
Pneumococcal meningitis, see Meningitis,
      pneumococcal
Pneumococcal pneumonia, see Pneumonia,
      pneumococcal
Pneumococcal sepsis, see Sepsis, pneumococcal
Pneumococci, see also *Streptococcus pneumoniae*
   complement activation and opsonization, 57,
      65
   serum sensitivity of, 141
Pneumonia
   cytomegalovirus, 100
   pneumococcal, 63—64
      C3 deficiency and, 32, 33
      complement activation in, 142
      immune complex nephritis, 142
Pneumonitis, with tropical pulmonary
      eosinophilia, 133
Polyarteritis nodisa, hepatitis B and, 107
Polymorphonuclear leukocytes, 8, 15
   C3b receptors of, 3
   chemotaxis, 17
   complement activation and, 16
   complement independent phagocytosis by, 10
   Streptococcus pneumoniae, 7
   streptococcal M protein and, 62
Polymorphs, *Treponema pallidum* adherence to,
      14
Polyoma virus, 97
Polysaccharide
   cryptococcal, 87, 88
   endotoxin molecule, 71
   fungal, 85
   pneumococcal, 63, 64
PPD, see Purified protein derivative
Properdin 1, see also Alternative pathway
   alternative pathway, 4, 5
   pneumococcal pneumonia and, 63
   Pseudomonas aeruginosa opsonization, 47
Propionibacteria, 66
   acne vulgaris, 144
   complement activation and opsonization, 56
   immune complex nephritis, 142
Protein(s), see also Antigens
   PPD, 67
   staphylococcal A, 59, 60, 99
   streptococcal, M, 61, 62
   viral, 93
Protein-calorie malnutrition, 38
Proteus mirabilis, 52
Proteus vulgaris, C3 deficiency and, 32
Protoscolices, echinococcal, 127
Protozoa, 115—125
   antigenic variation of, 115
   habesia, 119
   classes of, 116
   Entamoeba histolytica, 124
   leishmania, 123—124
   plasmodia, 116—119
   trypanosomes, 120—123

Pseudomonas aeruginosa, 47—48
   complement deficiencies and, 32, 33
   opsonization of, 12
Purified protein derivative, *Mycobacterium
      tuberculosis*, 67

R

Rabies virus, 97
Receptors, C3, 14—15
   biological roles of, 6—16
   B lymphocytes, 14—15
   Epstein-Barr virus receptors, 100
   erythrocytes, 14
   intraphagocytic microbe killing, 13—14
   opsonization, 6—13
   platelet, 18
   renal glomeruli, 15
   variability among, 12—13, 15—16
Receptors, C3b, 3
Receptor, Fc, 4, 11
Reticuloendothelial system, 8
Retroviruses, 97, 105—106
Rhabdoviruses, 105
Rhinotracheitis virus, infectious bovine, 101
Rhizopus, 86
Rickettsia rickettsii, 68
Rocky Mountain spotted fever, 68
Roundworms, 117, 132—133
Rous fowl sarcoma virus, 105
Rubella virus, 103

S

Sabin-Feldman dyte test, 124
Salmonellae, 50—51
   classical pathway deficiencies and, 31
   O side chains, 45
Salmonella enteridis, 50
Salmonella minnesota, 50
Salmonella typhimurium, 50—51
   capsule of, 58
   lysis of, 19
   sickle cell disease, 37
Schistosoma haematobium
   immune complexes with, 128
   in man, 127
Schistosoma japonicum
   in hamsters, 128
   in man, 127
Schistosoma mansoni, 128
   immune complexes with, 128
   in man, 127
Schistosomes
   concomitant immunity, 128—129
   host defenses, 129
   immune complexes, 128, 142
   leukocytes and, 131—132
   life cycle of, 127—128

serum sensitivity of, 129—132, 142
species pathogenic for man, 127
Schistosomula, serum sensitivity of, 129—131
Schwartzmann reaction, endotoxins and, 73—74
Scolices, echinococcal, 127
Semliki forest virus, 97, 102
Sepsis
   complement activation, 142
   complement deficiencies and, 32, 34
   factor B activation, 141
   Gram-negative
      and complement activation, 142
      endotoxins and, 71
   Haemophilus influenzae, 52—53
   pneumococcal, 31, 32
Serratia marescens, 12, 52
Serum
   bacterial effect, 19
   complement inactivation, 22
Serum sensitivity
   bacteroides, 53
   Escherichia coli, 48—50
   microbes with, 141—142
   mycoplasma, 69
   Neisseria gonorrhoeae, 54—55
   Plasmodium knowlesi, 116
   salmonellae, 50—51
   schistosomiasis, 129—131
   Serratia marescens, 52
   shigellae, 51
   Trypanosoma cruzi, 123
   trypanosomes, 121
Serum sickness, 143—144
Shock
   dengue virus, 102—103, 141
   endotoxin, 71—72
Sialic acid, alternative pathway activators, 5
Sickle cell disease, 36—37
Sindbis virus, 97, 102
Smallpox, 97
South American blastomycosis, 86
Spirochetes, 46—47, 142
Splenectomy, 37—38
Sporotrichosis, 86
Staphylococcal protein A, 59
   and complement consumption, 60
   and herpes simplex neutralization, 99
Staphylococci, serum sensitivity of, 141
Staphylococcus albus, see *Staphylococcus epidermidis*
Staphylococcus aureus, 56—60
   C5 dysfunction and, 35
   capsules, 57—59
   cell walls of, 58—60
   complement activation and opsonization, 8, 12, 56—57
   complement deficiency and, 32
   immune complex disease, 144
   intracellular killing of, 13
   opsonization of, 8, 12, 56—57
   peptidoglycans, 58—60
Staphylococcus epidermidis, 55—56

antibody and opsonization of, 12
complement activation and opsonization, 56
immune complex nephritis, 142
opsonization of, 8
Streptococcal infections
   chronic, immune complexes in, 142
   pharyngitis, classical pathway deficiencies, 31
   serum sickness-like disorder, 143—144
Streptococci
   classification of, 60—61
   complement activation and opsonization, 61—62
   group B, 60, 61
Streptococcus bovis, 60, 61
Streptococcus faecalis, 60, 61
   C5 deficiency and, 33
   complement reactivity, 61
   endocarditis, 62
   opsonization of, 12
Streptococcus mutans
   complement reactivity, 61
   periodontal disease, 144
Streptococcus pneumoniae, 63—66
   classical pathway deficiencies, 31
   classification of, 63
   complement activation by, 63—64, 142
   complement deficiency and, 32, 33
   immune adherence, 14
   immune complex nephritis, 142
   intracellular killing of, 14
   phagocytosis of, 7
   pulmonary clearance of, 64
   sickle cell disease, 36, 37
Streptococcus pyogenes, 60, 61
   C3 deficiency and, 32
   classical pathway deficiencies, 31
   glomerulonephritis, 142, 144
   M protein deficient, 62
Streptococcus viridians, 60, 61
Strongyloidiasis, 117
Superoxide
   phagocytosis assay, 10
   production, 17
Synthesis, complement, 1
Syphilis, 46—47, 142

T

Taenia, 126
Tapeworms, 116, 126—132
Teichoic acids, 43—44
   host defenses and, 46
   staphylococcal, and complement activation, 58—59
   streptococcal, 61
Terminal comcomplement sequence, 6
   deficiencies of, 33—35
   intracellular killing, 13
Thai hemorrhagic fever, 107
β-Thalassemia, 37
Thermoactinomyces vulgaris, 68

Togaviruses, 102—103
Torulopsis glabrata, 86, 87
Toxocara infections, 117
Toxoplasma gondii, 116, 124—125
Trematodes, 117, 127—132
Treponema pallidum, 46—47
   erythrocyte adherence, 14
   immune complex nephritis, 142
Trichinella spiralis, 117, 132, 133
Trichomonas vaginalis, 116
Trichostrongylus species, 117
Trichuris trichiura, 117
Tropical pulmonary eosinophilia, 133
Trypanosoma congolense, 121—122
Trypanosoma cruzi, 120, 122—123
Trypanosoma cyclops, 122
Trypanosoma duttoni, 122
Trypanosoma lewisi, 122
Trypanosoma musculi, 122
Trypanosomes
   antigenic change, 115
   nonhuman, experimental infections with, 121—122
   serum sensitivity of, 141
Trypanosomiasis, African, 120—121, 142
T-strain mycoplasmas, 70
Tuberculosis, 67
Tularemia, 48
Typhoid fever, 51

U

Ureaplasma urealyticum, 70
Urinary tract infections
   C3 deficiency and, 32
   classical pathway deficiencies, 31
   mycoplasmal, 70
Urticatia, with Epstein-Barr virus infection, 100

V

Vaccinia virus, 97—98
Vasodilation, C3a and C5a and, 17

Ventriculoatrial shunt infections
   corynebacterial, 66
   immune complex nephritis, 142
   Listeria monocytogenes and, 66
   Propionibacterium acnes, 66
   Staphylococcus epidermidis and, 55
Vesicular stomatitis virus, 105
Vibrio cholera, lysis of, 19
Viridians streptococcus, 60—62
Viruses
   arena viruses, 106—107
   bacteriophages, 108—109
   coat characteristics, 96
   coronaviruses, 107
   hepatitis B virus, 107—108
   herpes viruses, 98—101
   immune complex disease, 141, 142
   interaction with host, 93, 95—97
   lymphocytic choriomeningitis virus, 144
   man and other animals, 94—95
   mechanisms of complement-mediated neutralization, 96—97
   neutralization of, 31
   orthomyxoviruses, 103—104
   papovaviruses, 101
   paramyxoviruses, 104—105
   picornaviruses, 101—102
   poxviruses, 97—98
   retroviruses, 105—106
   rhabdoviruses, 104
   serum sensitivity of, 141
   togaviruses, 102—103

W

Western equine encephalitis virus, 102
Whipworm infection, 117

Z

Zymosan
   and alternative pathway, 4
   and complement-coagulation interaction, 18